Optical Network Theory

The Artech House Optoelectronics Library

Brian Culshaw, Alan Rogers, and Henry Taylor, Series Editors

Acousto-optic Signal Processing: Fundamentals and Applications, Pankaj Das

Amorphous and Microcrystalline Semiconductor Devices, Optoelectronic Devices, Jerzy Kanicki, ed.

Highly Coherent Semiconductor Lasers, Motoichi Ohtsu

Optical Control of Microwave Devices, Rainee N. Simons

Optical Fiber Sensors, Volume I: Principles and Components, John Dakin and Brian Culshaw, eds.

Optical Fiber Sensors, Volume II: Systems and Applicatons, Brian Culshaw and John Dakin, eds.

Principles of Modern Optical Systems, Volume I, I. Andonovic and D. Uttamchandani, eds.

Principles of Modern Optical Systems, Volume II, I. Andonovic and D. Uttamchandani, eds.

Reliability and Degradation of LEDs and Semiconductor Lasers, Mitsuo Fukuda

Optical Network Theory

Yitzhak Weissman

Artech House
Boston • London

Library of Congress Cataloging-in-Publication Data

Weissman, Yitzhak.
Optical network theory / Yitzhak Weissman.
p. cm.
Includes bibliographical references and index.
ISBN 0-89006-509-8
1. Optical communication. I. Title.
TK5103.59.W45 1992 91-48218
621.382'7-dc20 CIP

685 Canton Street
Norwood, MA 02062

International Standard Book Number: 0-89006-509-8
Library of Congress Catalog Card Number: 91-48218

10 9 8 7 6 5 4 3 2 1

To my parents

Table of Contents

Preface

The idea that network theory concepts can be applied to the analysis of optical systems occurred to me sometime in the late eighties. In the beginning, I regarded this idea more as an anecdote rather than as a viable scientific concept. However, it somehow managed to settle down in my mind and initiate a process over which I gradually lost control. I started to refresh my education in electrical and microwave network theories, and I took a second look at the theory of stochastic processes. Step by step, the significance of the optical network theory concept began to unfold. In the summer of 1989 I took a sabbatical leave from Soreq and moved to the Electrical Engineering Department at the University of Utah. Relieved from all my professional duties, I became possessed by the optical network theory idea.

At that time, some key results were already established, but most of the material still waited to be unveiled. With the excellent working conditions provided by the university and the beautiful surroundings of Salt Lake City, my research progressed fast. I remember that several important breakthroughs occurred while hiking the spectacular trails of the Millcreek Canyon. I soon realized that this research would create enough material for a full-size book. At this stage I was lucky enough to find Artech House. After a few weeks of correspondence with me, this publisher was persuaded to gamble on an unknown writer writing on an unheard-of subject, and a contract was signed.

In November 1990 I returned to Soreq with about half of the book written down and most of the second half written in the form of research notes. One chapter in the second part of the book was written in Tel Aviv under the atrocious terror of ballistic missile raids. In spite of the far reaching effects that these raids had on routine life, they had little influence on my progress, since at this step I knew already that the project would be completed.

Scholarly ethics require a comprehensive literature review to accompany any research. I am sorry to admit that I did not comply with this rule, partly because this book covers subjects from several disciplines, only one of which I can claim to be familiar with. My citations are incomplete, sporadic, and biased by personal communications. I would like to apologize to all the workers whose contributions were unintentionally omitted.

Yitzhak Weissman
Tel Aviv, October 1991

Acknowledgments

Throughout my work on this book I have enjoyed the help and support of many individuals. My thanks start with my colleagues Ehud Shafir and Moshe Tur for numerous discussions and for introducing me to the fascinating subject of field-induced noise. Craig K. Rushforth, the Chairman of the Electrical Engineering Department at the University of Utah, generously provided me with all I needed to work on this project during my stay in his department. I also enjoyed help and encouragement from department faculty members Carl H. Durney and Douglas A. Christensen. Special thanks are due to the Soreq Library and Documentation Department staff for letting me use its facilities to produce the manuscript. I am also obliged to the anonymous Referee for reading the manuscript and making many useful suggestions. Parts of the manuscript were meticulously reviewed by my colleague Arthur Schoenberg.

The manuscript of this book has been prepared in a camera-ready form using Microsoft Word on the Macintosh Classic. Most of the drawings were done with MacDraw II. Pictures illustrating the behavior of various functions were prepared with the help of THINK's LightspeedC and Cricket Graph. I would like to salute the creators of these hardware and software products, which gave me reliable service and predictable performance throughout the project.

Finally, I want to thank my dear wife Maayana and children Iris and Guy for continuous support, in spite of the long periods of my physical and mental absence they had to put up with.

Part 1:

Network Analysis

Chapter 1
Introduction

1.1 WHY OPTICAL NETWORKS ARE DIFFERENT FROM MICROWAVE NETWORKS

In many respects, optical and microwave networks seem similar. In both cases, we are dealing with an assembly of components interconnected with waveguides, and the signals are narrowband electromagnetic fields. Many microwave components have optical analogs, and the list is growing every year. In fact, many engineers with a background in microwave technology entering the field of optics regard optical networks as an "extension" of microwave networks, both technologically and theoretically. At a first glance, the most striking difference between microwave and optical networks seems to be the fact that optical networks offer bandwidths orders of magnitude higher than microwave networks. At second glance, technological differences become apparent: the optical components and waveguides are dielectric, and some sizes are scaled down three to four orders of magnitude. In particular, the effective field diameter in single-mode optical waveguides is measured in microns, which leads to strict mechanical tolerances in splices and connectors. In Table 1.1 we list the frequencies of the most popular bands used for optical network applications.

It would be naive to believe that an increase of five orders of magnitude in the field frequency will pass as simply as that. Indeed, there is a whole range of optical phenomena that make optical networks different from microwave networks. One category of such phenomena is related to the field-material interaction. Due to the short wavelength of the optical fields, their interaction with the microscopic constituents of matter is not negligible. The scattering of light by the molecules of the waveguide material, called *Rayleigh scattering*, is a very familiar phenomenon which can be easily observed with commercial optical time-domain reflectometer (OTDR) instruments. Due to the intrinsic vibrations of electrons, atoms, and molecules, there are several inelastic scattering mechanisms that involve an exchange of energy between the matter and the field. Such a mechanism is, for instance, the ***Brillouin scattering mechanism***, which involves an energy exchange between the field and the molecular vibrations (phonons). These types of phenomena affect the properties of the waveguides and the network components.

Table 1.1 The Frequencies of the Optical Carriers in the Three Most Popular Optical Bands.

Wavelength (μ)	Frequency (THz)
0.85	353
1.31	229
1.55	194

There is also another category of phenomena which manifests itself on the network theory level. The common "single-mode" optical waveguides, like the single-mode optical fibers, are, strictly speaking, two-mode waveguides. The two propagating modes that they support are nearly degenerate and differ mainly by their polarization. Accordingly, a given linear combination of the two propagating modes is called in optics the *state of the polarization of the field.* Most optical components are sensitive to the state of the polarization, and, therefore, a full account of its evolution is essential in optical network analyses. As we will see, this implies that the optical network algebra is a matrix algebra. In particular, the transfer function in optics becomes a transfer matrix, and the product of such transfer matrices is in general noncommutative.

Another range of phenomena that sets a fundamental difference between optical and microwave networks stems from the signal properties. The signal that is generated by microwave sources is normally considered fully deterministic, or, in other words, ideally coherent. This is justified in view of the fact that the spectral extension of the noise that accompanies a synthesized microwave signal is usually below a few tens of hertz. Such a noise is normally of little concern, since it can be easily filtered out without affecting the information signal. Moreover, the microwave network transfer function can be assumed to be constant over a frequency band of a few tens of hertz. For these reasons, the statistical properties of synthesized microwave signals are normally not considered.

The noise that accompanies the generation of optical signals has a much larger spectral extension. The presence of this noise is evidenced by the existence of a finite source linewidth. The value of the linewidth may vary over many orders of magnitude, depending on the type of the source. The linewidths of several common optical sources are listed in Table 1.2.

As can be seen from Table 1.2, the optical source noise spectrum may overlap the information signal spectrum, leading to a reduction of the signal-to-noise ratio. Furthermore, the network transfer function may vary significantly over the signal noise linewidth. For a proper evaluation of the noise properties of the output signals, the second- and fourth-order statistical properties of the input signals must be considered.

Table 1.2 The linewidths and the coherence lengths of the optical sources that are used in optical network applications. The quoted values are only order-of-magnitude estimates, and the corresponding values for real devices may be significantly different.

Type of Source	Linewidth	Coherence Length
Light-emitting diode	~20 THz	~15μ
Superluminescent diode	~4 THz	~75μ
Single-mode laser diode	~1 GHz	~30 cm
Distributed-feedback laser diode	~50 MHz	~6 m
External-cavity laser diode	~50 KHz	~6 km

This state of affairs may be described alternatively as follows. The coherence length of synthesized microwave signals is enormous (hundreds of thousands of kilometers), and for all practical purposes may be regarded as infinite. On the other hand, the coherence length of common optical sources may be compatible with the network size and even with the size of the network components. For these reasons, effects associated with finite coherence length cannot be neglected in optical networks.

1.2 TO WHOM THIS BOOK IS ADDRESSED

This book is not about optical communication. It deals with the more fundamental aspects of optical networks, and in a way it presents an optical analog to the microwave network theory. We believe that the book will be useful for both communication and sensing applications.

Nowadays, the research and development in the field of optical networks is carried out by personnel coming from two rather different disciplines: optics and electrical engineering. Naturally, each discipline has its own heritage regarding nomenclature and methodology. We made an effort to make this book intelligible and useful for both groups by identifying the related concepts and names. Although the lack of adherence to one of the disciplines may create in the beginning some confusion, we believe that, in the long run, all readers will benefit from our approach, as the introduction of a uniform style will facilitate the accumulation and proliferation of knowledge in the field.

The book presents some new concepts and approaches. The utility of the results ranges from fundamental research to applied design. Thus, both academic researchers and optical network designers working for industry may find something of interest in the book. The subject of the book is *methodology* rather than analysis. We do not present in-depth analyses of optical networks, but rather a methodology of how to perform such an analysis. The numerous examples that appear in the book serve as illustrations of the use of this methodology rather then as real-life analyses. They were chosen primarily in view of their simplicity and pedagogical value. In many instances it may seem that the examples can be treated effectively with much simpler, commonly used adhoc methods, and our treatment appears to be "overkill." It must be borne in mind, however, that these examples illustrate the use of more powerful and general methods, which can be applied to much more complex problems, where it would be difficult to get around with the common methods.

1.3 BOOK OVERVIEW

The book is organized into two parts: Part 1 is devoted to network analysis, and Part 2 to signal analysis. Even though Part 2 uses some fundamental concepts introduced in Part 1, we believe that a reader with a sufficient background in optics can follow the exposition in Part 2 without reading Part 1 first. This belief is based on the fact that we have used the traditional terminology whenever possible; the meaning of the fundamental concepts can usually be figured out using common knowledge and intuition. Thus, although the two parts are not exactly independent, they may be treated as such by experienced readers.

1.3.1 Network Analysis

Part 1 consists of Chapters 2 through 7. Chapter 2 explains the basic concepts upon which much of the material is based, namely *Jones vectors* and *Jones matrices*. Although originally introduced for a different purpose, these concepts are very suitable for the description of guided fields and their transformations. The meaning of Jones matrices is extended, and it is shown that they may be regarded as the transfer matrices of network components. In the rest of the book, Jones and transfer matrices are regarded as synonyms. In this chapter we also lay the foundations for the future treatment of time-periodic networks.

Chapter 3 is an introduction to the utilization of the *S-matrix* for the characterization of optical components. In contrast to microwave components, we consider here components which can accept two independent signals (modes) at each port. To characterize such components, we propose to use matrix scattering parameters. The S-matrix of such components becomes therefore a "super" matrix; i.e., a matrix with elements which are 2×2 matrices themselves. Each element of this S-matrix is identified with the transfer matrix between the two corresponding ports. Once the S-matrix is properly defined, we may invoke the power conservation and reciprocity requirements, which impose certain general relations between the S-matrix elements. These relations are generally not satisfied by plain Jones matrices.

Chapter 4 explains the *signal flow graph method* and its application to optical networks. We show how a signal flow graph is derived from the physical scheme of the network, and present the basic graph reduction rules. These rules take into account the fact that the product of the branch transmissions is noncommutative. The main purpose of this chapter is to provide general tools for the derivation of the network transfer matrix. The algorithms are presented with an abstract algebra. This allows us to apply them in both the time-dependent and the time-independent cases, which differ in their algebras.

Chapter 5 deals with the derivation of the transfer matrix for time-independent networks. We start by presenting the general algorithms and then illustrate our approach by analyzing four simple and common optical networks. These networks are the *Fabry-Perot interferometer*, the *recirculating loop*, the *Mach-Zehnder interferometer*, and the *Michelson interferometer*. We will also use these networks in subsequent illustrations throughout the book.

The analogous treatment of time-periodic networks is presented in Chapter 6. The algebra of the time-periodic networks is explained in detail because it is considerably different from its time-independent version. As an example, we treat first the problem of how to combine several modulators. Then we analyze two simple network applications: the cases of the modulated recirculating loop and the modulated Mach-Zehnder interferometer.

Perhaps the most sophisticated example of an optical network is the *fiber-optic gyro*. In view of the great interest in this system, we have decided to devote to it a special chapter. The material presented in Chapter 7 is an introduction to the fiber-optic gyro theory, and it also serves as an advanced illustration of the new methodology. In particular, this is one of the few instances in which polarization modes are considered explicitly. This chapter should not be considered an in-depth analysis of the fiber-optic gyro. Such an analysis is out of the scope of the book, and interested readers are referred to the literature. Chapter 7 concludes the first part of the book.

1.3.2 Signal Analysis

Optical signal analysis is based on the assumption that the optical field is a random process, and that the corresponding tools from the random processes theory can be invoked to compute the relevant quantities, like the power spectra. Although the random processes theory is well documented in the literature, we start the second part with a short review of the relevant fundamentals. This is done primarily for the convenience of the reader, since we do not introduce any new terminology or conventions as far as abstract random processes are concerned. Chapter 8 proceeds with the computation of optical quantities based on second-order field statistics.

The statistical properties or real optical sources may be quite complex. There are, however, two common models used in signal analysis: the *Gaussian model* and the *random-phase model*. These models are invoked to describe incoherent and coherent sources, respectively. The only source that is used in optical networks and for which Gaussian statistics is realistic is the light-emitting diode. The random-phase model is appropriate for the description of purely single-mode and intensity-stabilized laser diodes. Realistic descriptions of optical sources that lie between these two extremes can be synthesized by taking various superpositions of random-phase and Gaussian fields. Since we deal with linear networks, the results for such composite fields can be inferred from the corresponding results of the basic fields. For this reason, we will not consider such composite fields. The Gaussian model is a well-known model in the context of the random processes theory. The random-phase model is less well known, and the last part of Chapter 8 is devoted to its presentation.

Chapter 9 deals with the difficult subject of the *field intensity power spectrum*. This quantity is derived from the fourth-order statistical properties of the field, or, to be more specific, from the fourth-order field correlation functions. The novel feature in our treatment is the utilization of the frequency-domain representation. The advantage of this approach becomes evident when the effect of a network on an optical signal is being considered.

The frequency-domain representation of the fourth-order correlation functions of stationary processes contain some δ function singularities. To allow analytic manipulations with these functions, their singularities must be elucidated. The investigation of these singularities led us to what we called the analytic representation of the fourth-order correlation functions. The analytic representation has two useful features: the singularities are explicitly factored out, and the Gaussian component is separated from the nonseparable fourth-order core of the correlation function. The analytic representation allows the derivation of many useful results.

The field-induced noise is in general only one of the perhaps many noise mechanisms in the system. While many noise mechanisms are of electronic origin, there are two noise mechanisms that can be attributed to optics: the source-induced noise and the shot noise. It is not practical to compare the magnitude of the field-induced noise to the magnitude of the noise mechanisms of electronic origin. On the other hand, it is of considerable interest to compare the magnitudes of the field-induced noise and the shot noise. This comparison is presented in the concluding section of Chapter 9.

Chapter 10 is devoted to the investigation of the effect of optical networks on stationary optical signals, or, to be more specific, with the question: for a given source and a given network, what will be the output field and its intensity power spectra? We start with the derivation of the general formula in the frequency domain. The general formula contains the nonseparable fourth-order core function. This function is normally unavailable, and therefore the general formula has limited direct practical usefulness. On the other hand, it is very useful in deriving results for special cases, or in dealing with special situations in which certain approximations can be invoked. An example of a successful application to a special case is the important case of a random-phase source coupled to a dispersive waveguide. Another important case that yields a simple result is the case of the Gaussian source, where the fourth-order core vanishes.

An important case that lends itself to a simple analytic treatment is the case of the frequency-periodic networks. Generally speaking, this case applies to the situation in which the field superpositions that occur in the various network junctions are uncorrelated (the "incoherent limit"). Another condition is that in the frequency interval of interest (which is the spectral extent of the source), the network transfer matrix depends on the frequency periodically. This condition occurs most commonly whenever the frequency dependence of the network transfer matrix comes from the transfer matrices of the waveguides only. For frequency-periodic networks, certain approximations can be applied, which reduce the general formula to a very simple form. Some of the results that we present here were derived earlier using a time-domain approach. The frequency-domain treatment allows us to unveil the full utility of the frequency-periodic network approximation. In the discussion of the frequency-periodic networks, we introduce several new concepts. These are the network characteristic matrices and the network noise factors.

Chapter 11 presents some analytic tools for the computation of the network characteristic matrices and noise factors. These tools are based on the complex contour integration method. The exposition is followed by three examples that illustrate how to apply these tools. Readers who are not interested in these techniques may skip this chapter.

The signal analysis that is presented in Chapters 8 to 11 applies to time-independent networks only. Time-dependent networks and particularly networks with a periodic time dependence (time-periodic networks) have many important applications. The periodic time dependence comes normally from optical modulators, which are used to impress the information signal on the optical carrier. Chapter 12 presents some general methods for the signal analysis in such networks. The analysis shows that the problem of the time-periodic network is equivalent to a set of time-independent network problems, which can be treated using the methods presented earlier. Although the equivalent set of the time-independent network problems is in general infinite, in most practical cases the solution of only a small number of time-independent problems is required.

In time-periodic networks there is a new parameter: the *network period*, or *network frequency*. Obviously, this parameter does not have an analog in time-independent networks. The analysis shows that the ratio between the source linewidth and the network frequency is an important parameter that affects the nature of the output optical signal. In particular, we show that the background optical intensity noise is minimized if the source linewidth is much smaller than the modulation frequency.

The book concludes with an application of the optical signal analysis to the coherent laser radar (lidar) problem. Obviously, a lidar is not a guided wave system, and this chapter demonstrates the fact that the formalism developed in this book can be used for the analysis of certain unguided systems as well, particularly interferometers. In the course of the analysis of the lidar system, several interesting and practically important facts are elucidated. We have shown that the heterodyne signal is superimposed on a field-induced noise background. The magnitude of the field-induced noise increases with range. Under normal operating conditions, the field-induced noise becomes the dominant noise mechanism when the round-trip distance to the target exceeds the coherence length of the source.

1.4 THE SCOPE OF THE BOOK

Our treatment is limited to optical signals only. It starts at an imaginary point on the source surface, where the optical signal is being injected into the input waveguide. It ends at another imaginary point located on the detector surface where the optical signal is being converted to an electrical signal. Electrical signals are not treated explicitly, although the conversion of the optical output to an electrical signal is in many respects an integral part of the network function. This is done in order to limit the scope of the book, and also because the optical detection process is well documented in the existing literature.

To be concrete, we have assumed throughout the book that the number of guided modes in the network is two. The matrix expressions are of course valid for any number of modes, and the generalization of the expressions containing the number of modes explicitly is straightforward. However, in practice the applicability of the formalism to networks that support more than two modes is quite limited. This is due to the well-known fact that in practice it is impossible to control intermodal interactions in multimode networks, thus excluding the possibility of describing such a network in terms of a deterministic transfer matrix.

A fundamental assumption with the whole treatment is that all network components are linear in the field amplitude. In other words, our theory is a linear theory. Nonlinear components play increasingly important roles in optical networks in performing logical functions and acting as "optical memories." The consideration of nonlinear effects is also important in high-performance systems, like the fiber-optic gyro and coherent communication systems.

In spite of the remarkable progress, optical network technology is still in its infancy. New technologies and new components continue to emerge every year. In this respect, the development of an optical network theory at this stage may be regarded as premature, since the ultimate technologies and applications are currently uncertain. There are many relevant subjects that were left out, such as multiport networks and higher order statistics. There is little doubt that with the advancement of the optical network technology and with the diversification of applications, new subjects, which we cannot foresee at present, will emerge. We hope that the theory presented in this book will provide a sound basis for future developments.

1.5 MATHEMATICAL NOTATIONS AND CONVENTIONS

The most frequently used convention is the one that we introduced for the distinction between time-domain and frequency-domain quantities. Let $\mathrm{F}(t)$ be a function of time t. The frequency-domain representation of $\mathrm{F}(t)$ is simply its Fourier transform $F(\nu)$:

$$F(\nu) = \int dt\, \mathrm{e}^{2\pi i\nu t}\mathrm{F}(t) \tag{1.1}$$

Conversely, $\mathrm{F}(t)$ can be derived from $F(\nu)$ by the *inverse* transform:

$$\mathrm{F}(t) = \int d\nu\, \mathrm{e}^{-2\pi i\nu t}F(\nu) \tag{1.2}$$

For the sake of symmetry, we regard $\mathrm{F}(t)$ and $F(\nu)$ as the time-domain and the frequency-domain representations, respectively, of a certain abstract object F, although it must be borne in mind that, in general, $\mathrm{F}(t)$ and $F(\nu)$ are completely different functions. In the mathematical treatment, we often use both the frequency-domain and the time-domain representations. To avoid the introduction of separate symbols and yet allow an easy distinction between the two representations, we indicate frequency-domain quantities with italic type, and time-domain quantities with roman type, as illustrated in (1.1) and (1.2). This convention allows us to use the same symbol for the two representations of a given object. To further reduce the risk of confusion, we always show the list of arguments following the symbol of a function. The symbols f, ν, and μ are used to denote frequencies, and the symbols t, s, and τ are used to denote times. Thus, it is also possible to deduce the representation being used by consulting the argument list.

In the mathematical presentations, we have used several notations and conventions widely used in mathematical texts. For the benefit of those readers who are not accustomed to such texts, we explain here those notations that, in our opinion, may be less familiar. Throughout this book, we use the symbol i to denote the square root of –1. In electrical engineering texts, the same object is denoted by j, since i is reserved to denote an electric current. Here, we do not deal with electrical currents at all, and, therefore, the use of i to denote $\sqrt{-1}$ does not lead to a conflict. The convention that we adopt here regarding the symbol i is used in mathematical and physical texts. In optics texts, both symbols are used, according to the preference of the author. The use of the symbol i in this book does not leave many doubts regarding the background of the present author.

In the mathematical analysis we rely heavily on matrix algebra and import several notations and conventions from this field. All arrays (matrices and vectors) are denoted by boldface symbols. The unit matrix is denoted by $\boldsymbol{I}$. We denote the hermitian conjugate of an array $\boldsymbol{W}$ by $\boldsymbol{W}^\dagger$:

$$(\boldsymbol{W}^\dagger)_{ij} = (\boldsymbol{W}_{ji})^* \tag{1.3}$$

where the asterisk denotes complex conjugation. The transposition of a matrix $\boldsymbol{M}$ is denoted by $\boldsymbol{M}^T$:

$$(\boldsymbol{M}^T)_{ij} = \boldsymbol{M}_{ji} \tag{1.4}$$

A well-known algebraic rule relates the hermitian conjugate of a matrix product to the product of hermitian conjugated matrices:

$$(\boldsymbol{AB})^\dagger = \boldsymbol{B}^\dagger\boldsymbol{A}^\dagger$$

Similarly,

$$(\boldsymbol{AB})^T = \boldsymbol{B}^T\boldsymbol{A}^T$$

We do not separate the indices of a matrix by commas unless an index expression is used. Thus, to denote the element $(i + j,k)$ of a matrix $\boldsymbol{M}$, we use the notation $\boldsymbol{M}_{i+j,k}$.

A vector is regarded as a column matrix. Thus, if $\boldsymbol{A}$ is a vector, then

$$\boldsymbol{A} = \begin{pmatrix} A_1 \\ A_2 \\ \cdot \\ \cdot \\ \cdot \\ A_N \end{pmatrix} \tag{1.5}$$

and

$$A^{\dagger} = (A_1^*, A_2^*, \ldots, A_N^*) \qquad (1.6)$$

The modulus square $|A|^2$ of a vector $\boldsymbol{A}$ is the sum of the squares of the absolute values of its elements:

$$|A|^2 = \sum_j |A_j|^2 \qquad (1.7)$$

According to the matrix multiplication rule, we can also write

$$|A|^2 = \boldsymbol{A}^{\dagger}\boldsymbol{A} \qquad (1.8)$$

Occasionally, we also use the notation $|\boldsymbol{M}|^2$ when $\boldsymbol{M}$ is a matrix. In this case, $|\boldsymbol{M}|^2$ stands simply for $\boldsymbol{M}^{\dagger}\boldsymbol{M}$. The product $\boldsymbol{A}\boldsymbol{A}^{\dagger}$ is a matrix:

$$(\boldsymbol{A}\boldsymbol{A}^{\dagger})_{ij} = A_i A_j^* \qquad (1.9)$$

In some contexts, $\boldsymbol{A}\boldsymbol{A}^{\dagger}$ is called "the outer product" of the vector $\boldsymbol{A}$ with itself. A dot will be used to denote the scalar product of two vectors:

$$\boldsymbol{A} \cdot \boldsymbol{B} = \sum_j A_j B_j \qquad (1.10)$$

Another notation from matrix algebra that we use is $\mathrm{Tr}(\boldsymbol{M})$ to denote the trace of a matrix $\boldsymbol{M}$:

$$\mathrm{Tr}(\boldsymbol{M}) = \sum_i M_{ii} \qquad (1.11)$$

The matrix product inside the trace operation is commutative, i.e.,

$$\mathrm{Tr}(\boldsymbol{M}\boldsymbol{N}) = \mathrm{Tr}(\boldsymbol{N}\boldsymbol{M}) \qquad (1.12)$$

This theorem is used several times in the mathematical analysis. Also, we will often encounter multiple integrals. To simplify the presentation of such integrals, we use the following notation for a product of differentials:

$$d^N x = dx_1 dx_2 \ldots dx_N \qquad (1.13)$$

The limits of all integrals extend to infinity, unless noted otherwise. Summations over Jones vectors component indices include only two index values. Other summations extend over all integers.

This concludes the list of the general notations and conventions. More specialized notations are introduced and defined in the text; for example, $\langle F \rangle$ denotes the ensemble average of F.

1.6 LIST OF PRINCIPAL SYMBOLS

Below we present an alphabetical list of the principal symbols that appear in the book. In the brackets we show the physical units of the symbol. In this book we adopt a convention in which the Jones vectors are dimensionless, whereas the measurable quantities like intensity and power spectra have appropriate units. In the equations that relate Jones vectors or correlation functions to measurable quantities, it is necessary to use a power unit, which is denoted by ζ. In the second part of the book, we assume that the numerical value of this constant is one, and it is omitted in order to simplify the presentation. In order to obtain the correct units, it is necessary to reinsert this constant in the appropriate places with the appropriate power. We also show (wherever meaningful) in parentheses the first numbered equation in which the symbol appears, or, alternatively, the closest numbered equation to the point of its introduction.

Some of the symbols listed here may appear in the text with indices, primes, etc. In the case of functions, often both the time-domain and the frequency-domain representations are used. Only the time-domain representation of such symbols is listed. In the list, "W" stands for watt, "s" stands for second, "m" stands for meter, "Hz" stands for hertz, and "A" stands for amperes. "EGF" stands for equivalent Gaussian field.

$\boldsymbol{A}, \boldsymbol{B}$:	Jones vectors (2.1)
B:	Electrical bandwidth of a detector [Hz]
c:	Light velocity in vacuum [m/s] (2.7)
$\boldsymbol{C}$:	Jones vector of a source (3.1)
f:	Radio frequency [Hz]
f_c:	Field linewidth [Hz]
f_0:	Modulation frequency [Hz] (2.20)
$\boldsymbol{F}$:	Jones matrix of a waveguide (2.6)
$\mathrm{G}(t_1,t_2)$:	Coherency matrix (8.25)
$\mathrm{G}_{nmkj}(t_1,t_2,t_3,t_4)$:	Fourth-order field correlation function (9.1)
$\boldsymbol{H}(\nu_1,\nu_2)$:	H matrix (10.4)
$H(\nu_1,\nu_2)$:	H function (10.21)
$\langle \mathrm{I}(t) \rangle$:	Average instantaneous intensity [W] (8.32)
$\boldsymbol{I}$:	Unity matrix (2.5)

I:	The letter "I" is often used to denote the value of an integral
$j_n(\nu)$:	Fourier expansion coefficient of a cyclic Jones matrix (2.20)
$\mathbf{J}(t_1,t_2)$:	Jones matrix, transfer matrix (2.4)
k:	Propagation constant [m^{-1}] (2.1)
$\boldsymbol{K}(f)$:	K characteristic matrix (10.57)
$\boldsymbol{L}(f)$:	L characteristic matrix (10.58)
$N(f)$:	Optical intensity power spectrum [W^2/Hz] (9.47)
NF:	Noise figure (12.45)
$N^{(G)}(f)$:	Intensity noise power spectrum of the EGF [W^2/Hz] (10.56)
$NNF^{(G)}(f)$:	Network noise factor (10.66)
$NNF^{(rp)}(f)$:	Network noise factor (10.67)
$N^{(S)}{}_I$:	Shot noise equivalent optical intensity noise [W^2/Hz] (9.68)
NVC:	Network variance coefficient (10.76)
$\boldsymbol{P}$:	Jones matrix of a polarizer (7.8)
$P(\nu)$:	Field power spectrum [W/Hz] (8.9)
$\mathrm{R}(t)$:	Single-argument correlation function (8.8)
$\mathbf{R}(t)$:	Single-argument correlation matrix (8.30)
$RIN(f)$:	Relative intensity noise [Hz^{-1}] (9.65)
$RIN^{(G)}(f)$:	Relative intensity noise of the EGF [Hz^{-1}] (10.70)
RIV:	Relative intensity variance (10.73)
$RIV^{(G)}$:	Relative intensity variance of the EGF (10.74)
$\boldsymbol{S}$:	S-matrix (3.1)
$S(f)$:	Optical intensity power spectrum [W^2/Hz] (9.39)
t :	Time [s]
$\boldsymbol{t}$:	Branch transmission
T:	Period of a periodic component [s] (2.14)
T:	Common delay time of a network [s] (10.44)
$T(\nu)$:	Network power transfer function (10.6)
Δf:	Frequency period of frequency-periodic networks (10.44)
$\Delta G_{nmkj}(t_1,t_2,t_3,t_4)$:	Auxiliary correlation function (9.10)
$\Delta R_{nmkj}(t_1,t_2,t_3)$:	Three-argument auxiliary correlation function (9.12)
γ:	Phase modulation index (2.32)
γ:	Phase diffusion rate [Hz] (8.71)
Γ:	Complex phase modulation index (6.18)
$\Gamma_{nmkj}(t_1,t_2,t_3,t_4)$:	Gaussian field fourth-order correlation function (9.2)
η:	Detector responsivity [A/W]
$\Lambda(t)$:	Phase structure function (8.60)
ν:	Optical frequency [Hz]
τ:	Time [s]
ζ:	Power unit [W] (2.3)

Chapter 2
The Jones Calculus of Guided Fields

An electromagnetic plane wave in free space with a given frequency and a given propagation direction may be represented as a linear combination of two orthogonal polarization modes. It is thus possible to represent such a wave with a two-component vector whose components are the coefficients of this combination [1]. This vector is referred to as the *Jones vector*. The utilization of the Jones vector leads to a corresponding representation of optical components using *Jones matrices* [1]. The description of optical components and plane waves and their interactions in terms of Jones matrices and Jones vectors is called in the literature *Jones calculus*.

For quantitative calculations with the Jones calculus, it is necessary to define a pair of orthogonal *polarization modes* to serve as a basis. In free space, the polarization modes of the plane waves are degenerate (i.e., they are characterized by identical propagation constants for a given frequency). Consequently, there is a certain degree of arbitrariness in the choice of the basis polarization modes, and therefore also in the definition of the Jones vector and the Jones matrix. For reasons that are both historical and practical, the two orthogonal planar polarization states are used commonly as a basis for the Jones calculus of the free-space plane waves.

Even though the electromagnetic fields of guided modes do not necessarily exhibit a strictly defined polarization, in the fundamental propagation modes a certain dominant polarization feature is usually present, and it is therefore still appropriate to refer to them as "polarization modes." This dominant polarization feature is most commonly planar, and, therefore, the guided polarization modes and their corresponding Jones vectors can be regarded, to a certain extent, as the analogs of the free-space planar polarization modes and their corresponding Jones vectors. Another significant difference between the free space and the guided propagation phenomena is the fact that the polarization modes in the guided case are in general nondegenerate. This is the case, for instance, in integrated-optics waveguides, as well as in a special category of optical fibers known as polarization-preserving fibers. Therefore, in the Jones calculus of guided fields the arbitrariness associated with the choice of the basis polarization modes is usually absent.

The Jones calculus is very convenient for the treatment of guided-wave optical systems. In the analysis of such systems we are not concerned usually with the field distribution in the plane transverse to the direction of the waveguide; instead, we focus our attention on the field evolution in the propagation direction. The Jones calculus is an elegant way to eliminate the transverse field distributions from the analysis. Concepts analogous to the Jones vector have also evolved in the theory of microwave networks. In that context, however, only one propagation mode is normally considered. This is due to the fact that in many applications the microwave waveguides support strictly a single propagation mode. Furthermore, even in cases in which circularly symmetric waveguides (like coaxial cables) are used, where each propagation mode is at least doubly degenerate, effects associated with modal dependencies are usually far less important than in optics. Therefore, the microwave Jones vector analog, which is usually called the "complex wave amplitude," is taken to be a complex number (a scalar) rather then a two-component vector.

In addition to the adaptation of the Jones formulation to guided fields, we present below two generalizations that are of great importance to the further development of the optical network theory. The first one is the introduction of the time-domain Jones matrix and the identification of its Fourier transform with the original Jones matrix [2]. This concept will allow us to treat general guided fields, i.e., guided fields that are not necessarily monochromatic. The second generalization is associated with the treatment of time-periodic components (modulators). Such components play an important role in modern optical networks, and, therefore, any purely time-independent network theory formulation would have been of limited use.

2.1 THE JONES VECTORS FOR GUIDED FIELDS

Let us denote the electric field distributions of the guided modes in the transverse plane with $\boldsymbol{u}_1$ and $\boldsymbol{u}_2$, and take the z-axis to be parallel to the translational symmetry direction of the waveguide. We implicitly assume that there are only two guided modes. The electric field distribution **E** of a purely monochromatic guided wave with frequency ν can be cast generally in the form

$$\mathbf{E} = \{\boldsymbol{u}_1[a_1\exp(ik_1z) + b_1\exp(-ik_1z)] + \boldsymbol{u}_2[a_2\exp(ik_2z) + b_2\exp(-ik_2z)]\}e^{-2\pi i\nu t} \quad (2.1)$$

Equation (2.1) describes two counterpropagating waves. The coefficients a_1 and a_2 characterize a wave propagating in the positive z-direction, and the coefficients b_1 and b_2 a wave propagating in the negative z-direction. Correspondingly, two Jones vectors **A** and **B** will be associated with **E**:

$$\mathbf{A} = \begin{pmatrix} a_1 \exp(ik_1 z) \\ \\ a_2 \exp(ik_2 z) \end{pmatrix} e^{-2\pi i \nu t}, \qquad \mathbf{B} = \begin{pmatrix} b_1 \exp(-ik_1 z) \\ \\ b_2 \exp(-ik_2 z) \end{pmatrix} e^{-2\pi i \nu t} \tag{2.2}$$

For the most part, we will be interested in fields propagating in a given direction, and we will characterize them with a single Jones vector.

Through the years, different definitions of the Jones vector appeared in the literature; in particular, quite often the Jones vector is normalized in some fashion and the exponential time dependence is either factored out or suppressed. Here, we adhere to the original definition [1] in which the Jones vector is regarded as time-dependent and unnormalized. We also note that the vectors $\boldsymbol{u}_1$ and $\boldsymbol{u}_2$, and therefore also $\mathbf{E}$, have three components. This is not to be confused with the fact that the Jones vectors have two components only.

It is convenient to normalize the mode field distributions $\boldsymbol{u}_1$ and $\boldsymbol{u}_2$ in such a way that they carry a unit power ζ. In this convention, the powers P_p and P_n, flowing in the positive and the negative z-directions, respectively, are given by

$$P_p = |\mathbf{A}|^2 \zeta, \quad P_n = |\mathbf{B}|^2 \zeta \tag{2.3}$$

and the Jones vectors **A** and **B** are dimensionless.

As indicated explicitly in (2.1) and (2.2), the two propagation modes have, in general, different propagation constants, k_1 and k_2. In an optical context, this fact is usually referred to as *birefringence*. In a birefringent waveguide, the relative phase of the two Jones vector components vary with z, generally causing a change in the field distribution (polarization state) along the waveguide. There is, however, one important case in which the two propagation constants are equal: the case of the common single-mode fibers. Because of the circular symmetry of these fibers, their two polarization modes are degenerate, and ideally the field distribution is preserved, regardless of the values of the Jones vector components. To comply with the free-space formulation, we will use the planar polarization modes as the basis for the Jones calculus in degenerate waveguides, although it may be argued that in circularly symmetric waveguides the two circularly polarized basis states are a more appropriate choice.

In a perfectly circularly symmetric fiber it is possible to excite circular, and even elliptic polarization modes in a complete analogy to the free-space situation. In practice, however, the state of polarization is not preserved. Unavoidable inhomogeneities and deviations from a perfect circular symmetry, as well as perturbations induced by external conditions cause the polarization state to evolve, usually in an uncontrolled manner. To prevent this from happening, special polarization-preserving fibers have been developed, which have the capability to preserve certain polarization states [3]. Nevertheless, it is often assumed that the state of polarization is conserved in a common fiber as well. It must be borne in mind that this assumption is valid only when the fiber in question is short enough (depending on its quality) and well protected from external perturbations.

2.2 JONES MATRICES

Originally, the Jones matrix $\boldsymbol{J}$ was introduced to represent a linear transformation of the Jones vector of a purely monochromatic field. Using matrix notation, such a transformation can be generally expressed as

$$\mathbf{A}' = \boldsymbol{J}\mathbf{A} \tag{2.4}$$

A Jones vector transformation may be induced by an optical component, as shown schematically in Figure 2.1. In this case, we refer to the Jones vectors **A** and **A'** as the input and output Jones vectors, respectively.

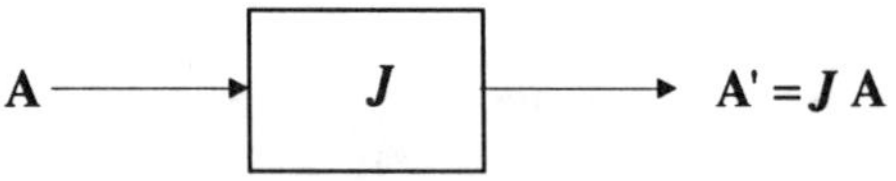

Figure 2.1 The transformation of a Jones vector.

The Jones matrix $\boldsymbol{J}$ is a 2×2 complex matrix. In general, its matrix elements are independent; consequently, eight independent parameters are necessary to completely specify a general Jones matrix. In practice, it is often assumed that the power is conserved, i.e., that $|\mathbf{A}'|^2 = |\mathbf{A}|^2$ for any **A**. This is equivalent to assuming that $\boldsymbol{J}$ is unitary:

$$\boldsymbol{J}\boldsymbol{J}^{\dagger} = \boldsymbol{I} \tag{2.5}$$

This equation stands for the following set of equations:

$$|J_{11}|^2 + |J_{12}|^2 = |J_{21}|^2 + |J_{22}|^2 = 1$$

$$J_{11}J^*_{21} + J_{12}J^*_{22} = 0$$

This set is equivalent to four real equations. Consequently, the Jones matrix of a power-conserving component contains only four independent parameters (one of which may be regarded as a common phase and is usually irrelevant).

Quite often, we will encounter components with diagonal Jones matrices. Such components do not couple the two polarization modes, or, in other words, if the input Jones vector has a vanishing component, the corresponding component in the output

Jones vector will vanish as well. We will call such components *polarization-preserving*. A smaller class of optical components have Jones matrices of the form JI, where J is some complex number. Such Jones matrices will be called *scalar*, or *degenerate Jones matrices*, since they can be treated in many respects as complex numbers.

Perhaps the most common component in optical networks is the *waveguide*. If **A** and **A'** denote the input and the output field Jones vectors, respectively, for a waveguide of length L, then we have

$$A'_1 = A_1 \exp(ik_1 L), \qquad A'_2 = A_2 \exp(ik_2 L)$$

Consequently, the Jones matrix $\boldsymbol{F}$ of such a waveguide is given by

$$\boldsymbol{F} = \begin{pmatrix} \exp(ik_1 L) & 0 \\ 0 & \exp(ik_2 L) \end{pmatrix} \tag{2.6}$$

The propagation constants k_1 and k_2 are related to frequency ν as follows:

$$k_1 = \frac{2\pi n_1 \nu}{c}, \qquad k_2 = \frac{2\pi n_2 \nu}{c}$$

where n_1 and n_2 are the effective refraction indices corresponding to the two polarization modes, and c is the speed of light in free space. We note that $\boldsymbol{F}$ is frequency-dependent and unitary. For a degenerate waveguide in which $n_1 = n_2 = n$, $\boldsymbol{F}$ becomes a degenerate Jones matrix:

$$\boldsymbol{F} = \exp(\frac{2\pi i L n}{c}\nu)\boldsymbol{I} \tag{2.7}$$

There exist components, called *polarizers*, that eliminate one of the polarization modes [4]. The Jones matrix $\boldsymbol{P}_m$ of a polarizer that allows only one mode $\boldsymbol{u}_m$ to pass is given by

$$\boldsymbol{P}_\text{m} = \begin{pmatrix} \delta_{1,m} & 0 \\ 0 & \delta_{2,m} \end{pmatrix}, \qquad m = 1, 2 \tag{2.8}$$

It is seen that $\boldsymbol{P}_m$ is a nonunitary matrix. The Jones matrices of real polarizers are frequency dependent, and in practice they assume the form expressed in (2.8) only for some rather narrow band of frequencies.

2.3 THE JONES MATRIX AS A TRANSFER FUNCTION OF A LINEAR SYSTEM

The original Jones calculus formulation was designed to treat perfectly monochromatic light. To account for effects associated with nonmonochromatic light, we will need a more general formulation. This generalization is easily achieved by applying the fundamental concepts of the linear system theory.

A general guided field $\mathbf{E}(t)$ propagating in a given direction can be represented by a general time-dependent Jones vector $\mathbf{A}(t)$:

$$\mathbf{E}(t) = A_1(t)\boldsymbol{u}_1 + A_2(t)\boldsymbol{u}_2 \tag{2.9}$$

Although we do not restrict explicitly the time dependence of $\mathbf{A}$, by assuming that only the two fundamental modes $\boldsymbol{u}_1$ and $\boldsymbol{u}_2$ are present, we imply that its spectral content is contained in the frequency band limited from below by the cutoff frequency of these modes, and from above by the cutoff frequency of the next higher modes. Suppose now that the field in (2.9) is modified by a linear optical component, creating a new field characterized by a Jones vector $\mathbf{A}'(t)$. The input and the output Jones vectors are related in general by

$$\mathbf{A}'(t_2) = \int dt_1 \mathbf{J}(t_2,t_1)\mathbf{A}(t_1) \tag{2.10}$$

The matrix $\mathbf{J}(t_2,t_1)$ represents the impulse response of the component. For a time-independent component, $\mathbf{J}(t_2,t_1)$ depends only on the time difference $t_2 - t_1$. We will call such Jones matrices *time-invariant*. It is convenient to associate with time-invariant Jones matrices a single argument, so that from now on a time-invariant Jones matrix $\mathbf{J}(t_2,t_1)$ will be denoted by $\mathbf{J}(t_2 - t_1)$. For a time-invariant Jones matrix $\mathbf{J}(t)$, we introduce the *frequency-domain* Jones matrix $\boldsymbol{J}(\nu)$ by

$$\boldsymbol{J}(\nu) = \int dt\, e^{2\pi i \nu t}\mathbf{J}(t) \tag{2.11}$$

The matrix $\mathbf{J}(t)$ may be also referred to as the *time-domain* Jones matrix [2]. The operation on the right-hand side of (2.11) is just a Fourier transform; thus, $\mathbf{J}(t)$ and $\boldsymbol{J}(\nu)$ are a Fourier transform pair.

In a linear-system context, the Jones matrix $\boldsymbol{J}(\nu)$ would have been called the "frequency-response matrix" or the "transfer-function matrix" of the component. In what follows, we will use the term *transfer matrix* as a synonym of Jones matrix. The time-domain and the frequency-domain Jones matrices of ideally frequency-independent components are related by

$$\mathbf{J}(t) = \boldsymbol{J}\delta(t) \tag{2.12}$$

that is, these components have an instantaneous impulse response. Therefore, for a frequency-independent component,

$$\mathbf{A}'(t_2) = \int dt_1 \boldsymbol{J}\delta(t_2 - t_1)\mathbf{A}(t_1) = \boldsymbol{J}\mathbf{A}(t_2) \tag{2.13}$$

We will demonstrate now that the original Jones matrix as introduced by us in (2.4) coincides with the frequency-domain Jones matrix defined in (2.11). The Jones vector **A** of a purely monochromatic field of frequency ν can be represented as

$$\mathbf{A}(t) = \boldsymbol{a}e^{-2\pi i\nu t}$$

The transformation of this vector by a time-invariant Jones matrix **J** is given by

$$\mathbf{A}(t_2) = \int dt_1 \mathbf{J}(t_2 - t_1)\boldsymbol{a}\exp(-2\pi i\nu t_1) = \boldsymbol{J}(\nu)\boldsymbol{a}\exp(-2\pi i\nu t_2)$$

or

$$\mathbf{A}' = \boldsymbol{J}(\nu)\mathbf{A}$$

which is the original Jones vector transformation rule (2.4).

To treat more general cases, namely nonmonochromatic fields and finite-time response components, we also introduce the *frequency-domain* Jones vector $\boldsymbol{A}(\nu)$. This vector is associated with $\mathbf{A}(t)$ by the Fourier transform relation

$$\boldsymbol{A}(\nu) = \int dt\, e^{2\pi i\nu t}\mathbf{A}(t) \tag{2.14}$$

We will refer to $\mathbf{A}(t)$ as the *time-domain* Jones vector.

For a time-independent component, the integral in (2.10) becomes a convolution. Consequently, the frequency-domain representation of this equation reads

$$\boldsymbol{A}'(\nu) = \boldsymbol{J}(\nu)\boldsymbol{A}(\nu) \tag{2.15}$$

We see that the original Jones vector transformation rule of a matrix-vector product is valid in the frequency domain for general fields, provided that the component is time-independent.

To familiarize ourselves with the time-domain Jones matrices, let us consider for example the time-domain Jones matrix of a waveguide. The time-domain Jones matrix $\mathbf{F}(t)$ corresponding to a waveguide of length L is obtained by applying the inverse Fourier transform to (2.6):

$$\mathbf{F}(t) = \begin{pmatrix} \delta(t-\tau_1) & 0 \\ 0 & \delta(t-\tau_2) \end{pmatrix}$$

where

$$\tau_k = \frac{Ln_k}{c}, \quad k = 1, 2$$

In this derivation we have implicitly assumed that the effective refraction indices n_k are independent of the optical frequency ν. In practice, this assumption is not true, since, in general, optical waveguides are dispersive. Accounting for dispersion would smooth the sharp δ function time dependence of $\mathbf{F}(t)$. The above equation corresponds to the impulse response of an ideal (power conserving and nondispersive) delay line exhibiting a delay τ_1 for mode $\boldsymbol{u}_1$ and τ_2 for mode $\boldsymbol{u}_2$.

When a monochromatic Jones vector $\boldsymbol{A}$ is incident upon a time-independent component characterized by a Jones matrix $\boldsymbol{J}(\nu)$, the output power P_{out} is given by

$$P_{\text{out}} = |\boldsymbol{A}|^2\zeta = \boldsymbol{A}^{\dagger}|\boldsymbol{J}(\nu)|^2\boldsymbol{A}\,\zeta \qquad (2.16)$$

Consequently, we will refer to the matrix $|\boldsymbol{J}(\nu)|^2$ as the *power transfer matrix.*

2.4 THE FOURIER REPRESENTATION OF CYCLIC JONES MATRICES

As we have already stated, the time-domain Jones matrix $\mathbf{J}(t_2,t_1)$ of a time-independent component depends only on the difference $t_2 - t_1$. In other words,

$$\mathbf{J}(t_2,t_1) = \mathbf{J}(t_2 + T, t_1 + T) \qquad (2.17)$$

for any T. Suppose now that T is the period of a component that is periodic in time. The Jones matrix $\mathbf{J}$ of such a component satisfies a discrete version of (2.17), namely

$$\mathbf{J}(t_2,t_1) = \mathbf{J}(t_2 + kT,t_1 + kT) \tag{2.18}$$

for any integer k and a given T. We will refer to such Jones matrices as *cyclic*. Let us analyze the frequency-domain characteristics of such Jones matrices.

We start by introducing $\boldsymbol{J}(\nu_1,\nu_2)$ with the help of the relation

$$\mathbf{J}(t_2,t_1) = \int d^2\nu \boldsymbol{J}(\nu_2,\nu_1)\exp[-2\pi i(\nu_2 t_2 - \nu_1 t_1)] \tag{2.19}$$

Now

$$\mathbf{J}(t_2+kT,t_1+kT) = \int d^2\nu \boldsymbol{J}(\nu_2,\nu_1)\exp\{-2\pi i[\nu_2 t_2 - \nu_1 t_1 + kT(\nu_2 - \nu_1)]\}$$

From (2.18) it follows that the difference $\nu_2 - \nu_1$ must be a multiple of the period frequency $f_0 = 1/T$, which means that it is possible to expand $\boldsymbol{J}(\nu_2,\nu_1)$ as follows:

$$\boldsymbol{J}(\nu_2,\nu_1) = \sum_n \boldsymbol{J}_n(\nu_2,\nu_1)\delta(\nu_2 - \nu_1 - nf_0)$$

By using this expansion in (2.19), we obtain

$$\begin{aligned}\mathbf{J}(t_2,t_1) &= \sum_n \int d^2\nu \boldsymbol{J}_n(\nu_2,\nu_1)\delta(\nu_2 - \nu_1 - nf_0)\exp[-2\pi i(\nu_2 t_2 - \nu_1 t_1)] \\ &= \sum_n \int d\nu_1 \boldsymbol{J}_n(\nu_1+nf_0,\nu_1)\exp\{-2\pi i[(\nu_1+nf_0)t_2 - \nu_1 t_1]\}\end{aligned}$$

Introducing

$$\boldsymbol{j}_n(\nu) = \boldsymbol{J}_n(\nu+nf_0,\nu) \tag{2.20}$$

we get

$$\mathbf{J}(t_2,t_1) = \sum_n \exp(-2\pi i n f_0 t_2)\int d\nu \boldsymbol{j}_n(\nu)\exp[-2\pi i \nu(t_2 - t_1)] \tag{2.21}$$

We will refer to (2.21) as the *Fourier representation* of $\mathbf{J}(t_2,t_1)$ and to the $\boldsymbol{j}_n$'s as its *Fourier expansion coefficients*. The Fourier representation is simpler than the conventional Fourier transform (2.19), because it involves a discrete summation instead of one of the integrations. Even though the summation is, in principle, infinite, in most cases of interest it is rapidly convergent. The Fourier representation is also more convenient for analytical purposes, since the time periodicity is built in. We also note that the time-independent component with a Jones matrix $J(\nu)$ may be regarded as a special case of the periodic component, with

$$j_n(\nu) = J(\nu)\delta_{n,0} \tag{2.22}$$

The Fourier expansion coefficients $j_n(\nu)$ can be computed from $\mathbf{J}(t_2,t_1)$ directly by the formula

$$j_n(\nu) = \frac{1}{T}\int_0^T dt_2 \int dt_1 \exp\{2\pi i[nf_0t_2 + \nu(t_2 - t_1)]\}\mathbf{J}(t_2,t_1) \tag{2.23}$$

We would like to point out that there is a certain freedom in the choice of the frequency-domain representation of $\mathbf{J}(t_2,t_1)$. For instance, had we performed the integration over ν_1 instead over ν_2 in arriving at (2.21), we would have reached a different (though equivalent) representation, namely

$$\mathbf{J}(t_2,t_1) = \sum_n \exp(-2\pi i n f_0 t_1)\int d\nu j_n'(\nu)\exp[-2\pi i\nu(t_2 - t_1)]$$

where

$$j_n'(\nu) = J_n(\nu,\nu - nf_0) = j_n(\nu - nf_0)$$

There is no reason at this stage to prefer any one of these two equivalent representations. The first representation, however, leads to a network algebra that blends well with other conventions, and so, in our opinion, it is easier to use and to interpret. We will refer, therefore, to (2.21) as the *Fourier representation*.

For later use we also introduce the time-domain counterparts $\mathbf{j}_n(t)$ of the Fourier expansion coefficients $j_n(\nu)$:

$$\mathbf{j}_n(t) = \int d\nu j_n(\nu)\exp(-2\pi i\nu t)$$

In view of (2.21), a cyclic Jones matrix can be also represented as follows:

$$\mathbf{J}(t_2,t_1) = \sum_n \exp(-2\pi i n f_0 t_2)\mathbf{j}_n(t_2 - t_1) \tag{2.24}$$

Thus, a cyclic Jones matrix may be represented as a linear combination of time-invariant Jones matrices.

Cyclic Jones matrices will be used to describe modulators. Let us consider the Jones vector $\mathbf{A}'(t)$ resulting from the action of a cyclic Jones matrix. Using (2.10) and (2.21) we obtain

$$\mathbf{A}'(t_2) = \sum_n \exp(-2\pi i n f_0 t_2) \int d\nu dt_1 \mathbf{j}_n(\nu) \exp[-2\pi i \nu(t_2 - t_1)] \mathbf{A}(t_1)$$
$$= \sum_n \exp(-2\pi i n f_0 t_2) \int d\nu \mathbf{j}_n(\nu) A(\nu) \exp(-2\pi i \nu t_2)$$

Taking now the Fourier transform of both sides, we get

$$A'(\nu) = \sum_n j_n(\nu - nf_0) A(\nu - nf_0) \tag{2.25}$$

We see that a time-periodic component generally creates new frequency components in the optical field, which is characteristic of the modulation process. If the original field is monochromatic with frequency ν_0, then $|\mathbf{j}_n(\nu_0)|^2$ determines the power in the frequency component $\nu_0 + nf_0$. Thus, the matrices $|\mathbf{j}_n(\nu_0)|^2$ may be regarded in a certain sense as the analogs of the power transfer matrix defined in (2.16) for the time-independent case.

2.5 EXAMPLES OF CYCLIC JONES MATRICES

The simplest modulation operation may be described by the equation

$$\mathbf{A}'(t) = \mathbf{V}(t)\mathbf{A}(t)$$

where $\mathbf{V}(t)$ is a periodic matrix function. Such an equation describes the action of an *instantaneous modulator*. In order to derive the Jones matrix $\mathbf{J}(t_2,t_1)$ corresponding to such a modulator, we cast this relation in the form

$$\mathbf{A}'(t_2) = \int dt_1 \mathbf{V}(t_2)\delta(t_2 - t_1)\mathbf{A}(t_1)$$

Comparing this with (2.10), we find that

$$\mathbf{J}(t_2,t_1) = \mathbf{V}(t_2)\delta(t_2 - t_1) \tag{2.26}$$

It is seen that $\mathbf{J}(t_2,t_1)$ is indeed cyclic in the sense of (2.18). To find the Fourier expansion coefficients $\boldsymbol{j}_n$, we use (2.23):

$$j_n = \frac{1}{T}\int_0^T dt_2 \mathbf{V}(t_2)\exp(2\pi i n f_0 t_2) = \boldsymbol{v}_n \tag{2.27}$$

where the series $\boldsymbol{v}_n$ is simply the discrete Fourier transform of the periodic matrix $\mathbf{V}(t)$. We see that, in this case, the Jones matrix Fourier expansion coefficients are independent of the optical frequency ν, reflecting the fact that the component has an instantaneous response.

Let us consider now a few specific examples of cyclic transmissions representing instantaneous modulators. We start with an instantaneous amplitude modulator:

$$\mathbf{V}'(t) = \frac{V_0}{2}[1 + \cos(2\pi f_0 t + \phi)] \tag{2.28}$$

where V_0 is a constant matrix and ϕ is an arbitrary phase. For example, this kind of modulation can be achieved for a plane wave by using three polarizers, the first and the last being fixed and parallel, and the middle one rotating with a frequency $f_0/2$. In this case

$$V_0 = \begin{pmatrix} 1 & 0 \\ 0 & 0 \end{pmatrix}$$

For guided fields, this kind of modulation is not practical; therefore, (2.28) should be regarded in the present context as an illustration only. Using (2.27), we find that the Fourier expansion coefficients $\boldsymbol{j}_n$ for this case are given by

$$j_n = \frac{V_0}{2}(\delta_{n,0} + \frac{1}{2}e^{-in\phi}\delta_{|n|,1}) \tag{2.29}$$

We see that the Jones matrix of this amplitude modulator is associated with only three nonvanishing Fourier expansion coefficients.

Next we consider the case of an instantaneous frequency shifter. Ideally, such a device is represented by the following periodic matrix $\mathbf{V}(t)$:

$$\mathbf{V}(t) = V_0\exp(-2\pi i f_0 t) \tag{2.30}$$

Using (2.27) again, we find that the Fourier expansion coefficients $\boldsymbol{j}_n$ for this case are given by

$$j_n = V_0\delta_{n,1} \tag{2.31}$$

The Fourier representation of an ideal frequency shifter therefore involves only one nonvanishing component.

Let us consider now the case of the phase modulation. An instantaneous phase modulator is represented by the following periodic matrix $\mathbf{V}(t)$:

$$\mathbf{V}(t) = V_0\exp[-i\gamma\sin(2\pi f_0 t + \phi)] \tag{2.32}$$

where γ is a real constant representing the modulation index. To find the Fourier expansion coefficients j_n of this Jones matrix, we use the well-known integral representation of the Bessel functions [5]

$$J_n(z) = \frac{1}{2\pi}\int_{-\pi}^{\pi} d\theta\exp(-in\theta + iz\sin\theta) \tag{2.33}$$

where the functions $J_n(z)$ are the Bessel functions (not to be confused with the Jones matrices). Using this integral to evaluate (2.27) for this case, we find

$$j_n = e^{-in\phi}V_0 J_n(\gamma) \tag{2.34}$$

In contrast to the two former cases, we see that the phase modulation operation is associated with an infinite number of nonvanishing Fourier expansion coefficients j_n. However, for most practical purposes only a finite number of coefficients needs to be considered, as $J_n(x) \rightarrow 0$ when $|n| \rightarrow \infty$ for a real x.

Phase modulation in optics is accomplished by components capable of varying the effective optical path length. Since the effective optical path length is the product of the refraction index n and the physical path length L, it can be varied by varying either n or L or both. In bulk and integrated-optics components, electro-optical materials are employed, and the effective optical path length is varied by varying n with an external electric field. A schematic drawing of such a modulator is given in Figure 2.2. In fiber-optic components, the phase modulation is achieved by actually stretching the fiber, most commonly with a piezoelectric material, thus varying L, as shown in Figure 2.3.

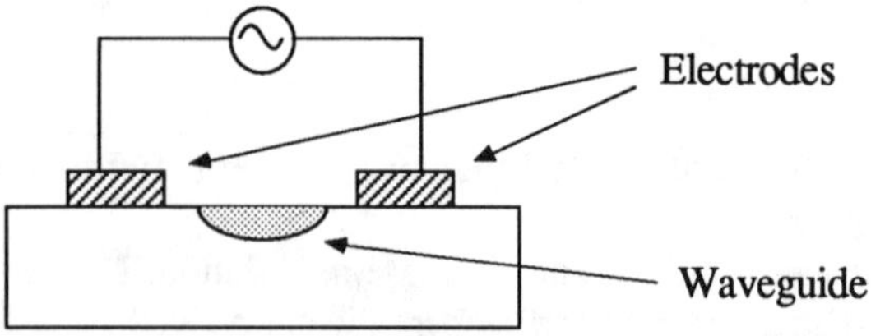

Figure 2.2 An integrated-optics phase modulator based on the electro-optic effect.

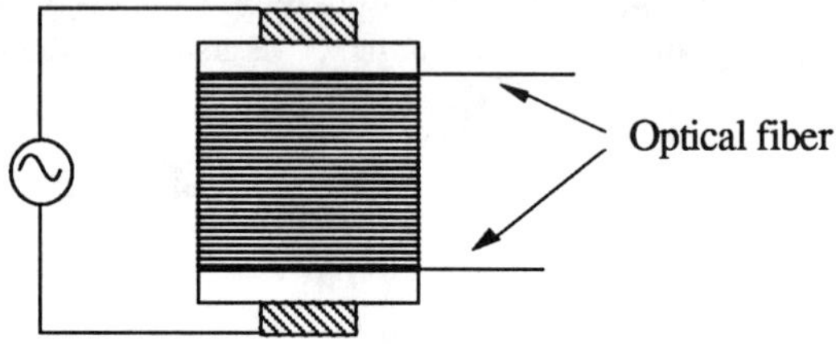

Figure 2.3 A fiber-optic phase modulator based on the mechanical stretching of the fiber.

Let us derive the Fourier expansion coefficients for the Jones matrix of these more realistic modulators. If the modulation period T is much larger then the light transit time through the device, we can assume that the output Jones vector **A**' is equal to the input Jones vector **A** at a certain time-dependent retarded point of time. However, since the retardation will, in general, be different for the two polarization modes, the input and the output Jones vectors are related as follows:

$$\mathrm{A}'_k(t) = \mathrm{A}_k[t - \tau_k + s_k \sin(2\pi f_0 t + \phi_k)], \quad k = 1, 2 \tag{2.35}$$

where τ_k and $s_k \sin(2\pi f t + \phi_k)$ are the average delay time and the time-dependent delay component for the kth mode, respectively. The Jones matrix $\mathbf{J}(t_2,t_1)$ corresponding to such a modulator is

$$\mathbf{J}(t_2,t_1) = \begin{pmatrix} \delta[t_2-t_1-\tau_1+s_1\sin(2\pi f_0 t_2+\phi_1)] & 0 \\ 0 & \delta[t_2-t_1-\tau_2+s_2\sin(2\pi f_0 t_2+\phi_2)] \end{pmatrix} \tag{2.36}$$

We leave it as an exercise for the reader to show that the Fourier expansion coefficients $j_n(\nu)$ corresponding to this case are given by

$$j_n(\nu) = \begin{pmatrix} J_n(2\pi\nu s_1)\exp(2\pi i\nu\tau_1 - in\phi_1) & 0 \\ 0 & J_n(2\pi\nu s_2)\exp(2\pi i\nu\tau_2 - in\phi_2) \end{pmatrix} \quad (2.37)$$

Note that the modulation index γ, which appeared as the argument of the Bessel functions in (2.31), has been replaced by $2\pi\nu s_k$. This reflects the fact that for a given amplitude of the time delay variation s, the amplitude of the phase modulation is equal to $2\pi\nu s$. In contrast to the three cases that we considered earlier, the Fourier expansion coefficients $\boldsymbol{j}_n$ in (2.37) depend on the optical frequency ν due to the fact that this phase modulator does not have an instantaneous time response.

2.6 SUMMARY

In this chapter we have introduced two important concepts: Jones matrices and Jones vectors. These concepts, introduced originally for the analysis of polarization transformations of optical beams, are presented here from the point of view of the linear system theory. In particular, we have shown that Jones matrices may be considered matrix transfer functions, and Jones vectors may be considered vector complex amplitudes.

Another subject we discussed in this chapter was time-periodic components. The Jones matrix of a time-periodic component is not invariant to an arbitrary time shift, like the Jones matrix of a time-independent component. Instead, it remains invariant only if its time arguments are shifted by a whole multiple of the time period of the component. Such Jones matrices were called cyclic. The frequency-domain representation of cyclic matrices is a linear superposition of an infinite number of time-invariant Jones matrices, which we have called Fourier expansion coefficients. As examples, we have presented the Fourier representations of a frequency shifter, an amplitude modulator, and a phase modulator.

References

[1] Jones, R.C., "A New Calculus for the Treatment of Optical Systems: I. Description and Discussion of the Calculus," *J. Opt. Soc. Am.*, Vol. 31, 1941, pp. 488-493.

[2] Parke, N.G., "Optical Algebra," *J. Math. Physics*, Vol. 28, 1949, pp. 131-139.

[3] Noda, J., K. Okamoto, and Y. Sasaki, "Polarization-Maintaining Fibers and Their Applications," IEEE *J. Lightwave Technol.*, Vol. LT-4, No. 8, 1986, pp. 1071-1089.

[4] Nishihara, H., M. Haruna, and T. Suhara, *Optical Integrated Circuits*, New York: McGraw-Hill, 1985, pp. 256-259.

[5] Rainville, E.D., *Special Functions*, New York: Chelsea, 1960, pp. 114.

Further Reading

[1] Shurcliff, W.A., *Polarized Light*, Cambridge, Massachusetts: Harvard University Press, 1962.

[2] Azzam R.M.A. and N.M. Bashara, *Ellipsometry and Polarized Light*, Amsterdam: North-Holland, 1989.

Chapter 3
S-Matrix Characterization of Optical Components

At the network level, we regard the components as "black boxes," isolated from the rest of the world except for a few designated *ports* that are accessible for external connection. For network analysis purposes, a component is completely characterized by the relations between the signals or the fields at these ports. This type of component characterization is called *port characterization.* In the frequency domain, the relations between the various fields at the ports of a linear and time-independent component are essentially a set of linear equations. Algebraically, the port characterization of a passive component can be represented by a matrix. For the characterization of active components, an extra vector describing signal generators may be needed.

Port characterization of components is very common in electrical and microwave networks analysis. This characterization is not unique and depends on the definition of the signals, or of the fields at the ports. In electrical networks, where current-voltage relations are of primary concern, impedance or admittance matrices are used, which relate the currents to the voltages at the ports. In microwave networks, where we are more concerned with travelling waves, the so-called *scattering parameters* are more useful. These parameters make up the *S-matrix* [1].

The S-matrix is also well suited as a port characterization of optical components. However, due to the fact that in the optical waveguides there are two guided modes, each physical port of an optical component is actually equivalent to two strictly single-mode ports. An optical component could be treated in exactly the same fashion as a microwave component by separating logically the two virtual single-mode ports that correspond to each one of the physical ports. However, this logical separation is inconvenient and confusing. Instead, we choose to regard each port both physically and logically as a single entity. This approach is more appealing conceptually, but there is a price to it: the scattering parameters become 2×2 matrices, and the S-matrix becomes a "super matrix," i.e., a matrix whose elements are 2×2 matrices instead of regular numbers. This is the most fundamental difference between the optical S-matrix as presented here and the corresponding S-matrix used in microwave theory.

In all other theoretical respects, there is a complete analogy between the S-matrix characterizations of optical and microwave components. In practice, however, there are some significant differences. These differences stem from the fact that optical wavelengths are on the order of a micron, three orders of magnitude smaller than millimeter waves. Thus, the S-matrix of optical components can be significantly modified by geometric distortions on the submicron scale. Such distortions can arise from thermal expansion or mounting stresses, which would have a negligible effect on microwave components. Certain optical components (couplers, for instance) have spatial dimensions that are much larger than the optical wavelength. In such components, the phases of the scattering parameters are extremely sensitive to temperature variations, since the effect of changes in the index of refraction on these phases is greatly amplified.

Most commonly, port characterization is done in the frequency domain and only for time-independent components. Since here we are also interested in time-dependent (or more specifically time-periodic) components, we will outline the proper generalization of the time-independent treatment to the time-dependent case. It must be remembered, though, that optical radiation can interact with the microscopic constituents of matter, and is therefore influenced by microscopic motion, such as molecular vibrations. This interaction can cause inelastic (Raman) scattering, namely, adding new frequency components to the signal even if the component is macroscopically time-independent. These new frequency components are offset by frequencies characteristic of the microscopic motion, which are considerably higher than RF or even microwave frequencies, and, therefore, do not normally interfere with the sidebands generated by the RF modulation of the optical carrier caused by macroscopic time variations. This fact normally allows us to ignore the new frequency components that arise from the inelastic scattering process, and to regard them as an effective loss. There may be cases in which this assumption is not valid; these cases, however, will not be considered here.

3.1 THE JONES MATRIX AND THE S-MATRIX (THE TIME-INDEPENDENT CASE)

Before introducing the formal definition of the S-matrix of an optical component, let us consider the simple case of a single-port component, as shown schematically in Figure 3.1. If the component is linear and time-independent, then in the frequency domain the input and the output Jones vectors $\boldsymbol{A}$ and $\boldsymbol{A'}$ are related by a linear set of equations

$$\boldsymbol{A'} = \boldsymbol{S}\boldsymbol{A} + \boldsymbol{C} \tag{3.1}$$

where $\boldsymbol{S}$ is a complex 2×2 matrix and $\boldsymbol{C}$ is a complex two-element vector, both independent of $\boldsymbol{A}$. The matrix $\boldsymbol{S}$ is called the S-matrix of the component. Drawing an analogy with the Jones vector transformation rule (Eq. (2.15)), we may also regard $\boldsymbol{S}$ as the Jones matrix of the component in this case. The Jones vector $\boldsymbol{C}$ represents optical field generators that may be present in the component. It has a significant value only in active components, such as optical sources or optical amplifiers. In passive components, $\boldsymbol{C}$ is normally neglected, since at the frequencies that are currently used in optical network applications (see Table 1.1), the blackbody radiation at room temperature is negligible. The neglect of $\boldsymbol{C}$ in passive components may not be permissible in the analysis of networks operating in the middle to far infrared.

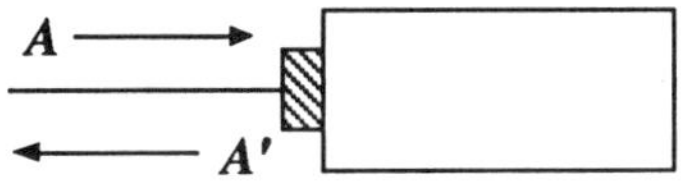

Figure 3.1 A scheme of a one-port component.

For the port characterization of components with a larger number of ports, a larger matrix is necessary. Consider now the situation shown in Figure 3.2, where the Jones vectors $\boldsymbol{A}_1, \boldsymbol{A}_2, ..., \boldsymbol{A}_N$ are incident upon an N-port component, so that $\boldsymbol{A}_k$ is the Jones vector incident upon port k. These Jones vectors give rise to another set of Jones vectors $\boldsymbol{A}'_1, \boldsymbol{A}'_2, ..., \boldsymbol{A}'_N$, so that $\boldsymbol{A}'_k$ emerges from port k. Let $\tilde{\boldsymbol{A}}$ and $\tilde{\boldsymbol{A}}'$ denote the total sets of the input and the output Jones vectors, respectively:

$$\tilde{\boldsymbol{A}} = \begin{pmatrix} \boldsymbol{A}_1 \\ \boldsymbol{A}_2 \\ \cdot \\ \cdot \\ \cdot \\ \boldsymbol{A}_N \end{pmatrix}, \qquad \tilde{\boldsymbol{A}}' = \begin{pmatrix} \boldsymbol{A}'_1 \\ \boldsymbol{A}'_2 \\ \cdot \\ \cdot \\ \cdot \\ \boldsymbol{A}'_N \end{pmatrix} \tag{3.2}$$

The vectors $\tilde{\boldsymbol{A}}$ and $\tilde{\boldsymbol{A}}'$ are "super vectors," i.e., vectors whose elements are the two component vectors $\boldsymbol{A}_k$ rather then simple numbers, so that their algebraic dimension is $2N$ rather then N. For a time-independent, linear component, the vectors $\tilde{\boldsymbol{A}}$ and $\tilde{\boldsymbol{A}}'$ are related by

$$\tilde{\boldsymbol{A}}' = \boldsymbol{S}\tilde{\boldsymbol{A}} + \tilde{\boldsymbol{C}} \tag{3.3}$$

This relation, which is a straightforward generalization of the single-port case considered above (3.1), may be regarded as a general definition of the S-matrix $\boldsymbol{S}$ of an N-port component.

In the same manner as we regard the Jones vectors $\tilde{\boldsymbol{A}}$ and $\tilde{\boldsymbol{A}}'$ as "super vectors," we logically treat $\boldsymbol{S}$ as a "super matrix," i.e., a matrix whose elements are

2×2 complex matrices. Accordingly, we will refer to the optical S-matrix elements as Jones matrices. This convention is again a straightforward generalization of the single-port case. However, from an algebraic point of view, $\boldsymbol{S}$ is a complex matrix of the order $2N$.

We can now put the contents of Chapter 2 in the proper perspective. An optical field can be transformed by feeding it into one of the ports of an optical component and collecting the output from any other port (including, in principle, the one used for input). Any such input-output combination is characterized by a Jones matrix. Since the number of possible input-output combinations in an N-port component is N^2, we conclude that N^2 Jones matrices are necessary to fully characterize a component. These Jones matrices are the elements of the S-matrix introduced above.

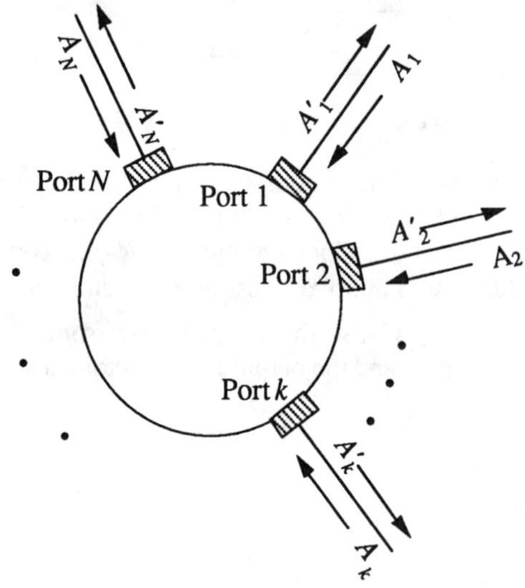

Figure 3.2 A scheme of an N-port component.

3.2 GENERAL PROPERTIES OF THE S-MATRIX

Some important properties of S-matrices are derived from energy conservation considerations [1]. The input and the output powers P_{in} and P_{out} are given by

$$P_{\text{in}} = |\tilde{A}|^2\zeta, \qquad P_{\text{out}} = |\tilde{A}'|^2\zeta = |\boldsymbol{S}\tilde{A} + \tilde{C}|^2\zeta \tag{3.4}$$

where ζ is a unit power introduced earlier (Eq. (2.3)). For a passive component, $\tilde{\boldsymbol{C}} = \boldsymbol{0}$, and energy conservation implies that $P_{in} \geqq P_{out}$, so that

$$\tilde{\boldsymbol{A}}^{\dagger}(\boldsymbol{I} - |\boldsymbol{S}|^2)\tilde{\boldsymbol{A}} \geqq 0$$

for any $\tilde{\boldsymbol{A}}$. Thus, for a passive component the matrix $\boldsymbol{I} - |\boldsymbol{S}|^2$ is semipositive. Components for which $P_{in} = P_{out}$ are called *power-preserving*, and accordingly their S-matrix is unitary:

$$|\boldsymbol{S}|^2 = \boldsymbol{I} \tag{3.5}$$

Power loss can occur as a result of absorption, inelastic scattering, and coupling to radiative modes.

Under normal conditions, optical components may be regarded as reciprocal. The reciprocity property is a fundamental symmetry relation that follows from the properties of the Maxwell equations. Reciprocal components are characterized by a symmetric S-matrix [1]:

$$\boldsymbol{S} = \boldsymbol{S}^T \tag{3.6}$$

The most common violations of reciprocity in optical components are due to magneto-optic interactions. As we will see later, the operation of the fiber-optic gyro is based on another source of nonreciprocity, namely, the Sagnac effect.

Certain optical components are designed to preserve the polarization state. The scattering parameters of such components are diagonal matrices, and there is no coupling between the two polarization modes. Consequently, such a component can be logically separated into two decoupled, strictly single-mode components, each being represented by a simple $N \times N$ S-*submatrix*. In particular, if the whole network is composed from polarization-preserving components (including the waveguides), then it can be logically separated into two strictly single-mode decoupled networks. In such networks the complication arising from the necessity to consider both polarization modes simultaneously is eliminated. Furthermore, we may imagine a situation in which a component, or even a whole network, is not only polarization-preserving, but also polarization-insensitive, meaning that its effect on both polarization states is identical. In this case, the two S-submatrices are identical, and the analysis may be completed by considering one polarization state only. We will call such components *degenerate components*. The assumption that a component is degenerate is, of course, only an approximation; this approximation is, however, very useful for preliminary and simplified analyses. We will use the degenerate component description in many of the examples for the sake of simplicity.

3.3 THE S-MATRICES OF SOME COMMON OPTICAL COMPONENTS

3.3.1 One-Port Components

The simplest one-port component is a *termination.* As discussed in Section 3.1, the S-matrix of a one-port component may be identified with its Jones matrix. There are two extreme cases of terminations: the matched termination and the perfectly reflecting, or mirror, termination. The S-matrix of an ideally matched termination vanishes, since ideal matching means that there is no reflected field for any incident field. Matched terminations are achieved in practice by immersing the bare end of the waveguide in an index-matched solution, or by polishing it at a certain angle. The mirror is the extreme opposite of the matched termination: the reflected power from an ideal mirror is equal to the incident power for any incident Jones vector. The S-matrix of an ideal mirror is therefore unitary. If, in addition, the mirror is also reciprocal (as is commonly the case), the S-matrix is also symmetric. Consequently, the S-matrix of a perfect and reciprocal mirror generally has the following form:

$$\boldsymbol{S} = e^{i\alpha}\begin{pmatrix} e^{i\beta}\cos\theta & i\sin\theta \\ i\sin\theta & e^{-i\beta}\cos\theta \end{pmatrix} \tag{3.7}$$

where α, β, and θ are three real parameters. The common phase $e^{i\alpha}$ does not normally have practical significance and is often suppressed. If $\theta = 0$ then the mirror is polarization preserving. If, in addition, $\beta = 0$ as well, then the mirror becomes degenerate.

The next single-port component that we discuss is the *source*. The general relation between the input and the output Jones vectors for single-port components is described in (3.1). For a source, the $\boldsymbol{C}$ vector does not vanish, reflecting the fact that optical power is being generated. For sources that generate their optical power by spontaneous emission, like light-emitting diodes, $\boldsymbol{S}$ may be considered an S-matrix of a single-port passive component; i.e., $\boldsymbol{I} - |\boldsymbol{S}|^2$ is semipositive. On the other hand, in laser sources the incoming field may penetrate the source cavity and undergo some degree of amplification, so that in this case it may not be true that $\boldsymbol{I} - |\boldsymbol{S}|^2$ is semipositive. However, $\boldsymbol{S}$ will normally be symmetric for both types of sources. In addition, experimental evidence points to the fact that in laser sources the vector $\boldsymbol{C}$ depends on the incoming Jones vector $\boldsymbol{A}$. Therefore, a laser source is inherently a nonlinear device. This nonlinear effect is usually undesirable in applications, and it is avoided by inserting an optical isolator between the source and the network. Since these isolators, in effect, absorb the fields incident on the source, we can describe a source-isolator combination simply by

$$\boldsymbol{A}' = \boldsymbol{C} \tag{3.8}$$

i.e., as a source with a vanishing S-matrix. This is indeed the common source description in optical network analyses.

3.3.2 Two-Port Components

The most common two-port component is the waveguide. The waveguides are the channels through which the fields propagate from one network component to the other. Ideally, these channels should be nonlossy and without reflections. Thus, the S-matrix of an ideal waveguide is unitary and has the form

$$\boldsymbol{S} = \begin{pmatrix} \boldsymbol{0} & \boldsymbol{F}_1 \\ \boldsymbol{F}_2 & \boldsymbol{0} \end{pmatrix} \tag{3.9}$$

where $\boldsymbol{0}$ stands for a 2×2 null matrix. Because $\boldsymbol{S}$ is unitary, it follows that $\boldsymbol{F}_1$ and $\boldsymbol{F}_2$ are unitary as well. Furthermore, if the waveguide is also reciprocal, then

$$\boldsymbol{F}_1 = \boldsymbol{F}_2^T$$

in which case the waveguide S-matrix can be written in terms of a single, unitary Jones matrix $\boldsymbol{F}$:

$$\boldsymbol{S} = \begin{pmatrix} \boldsymbol{0} & \boldsymbol{F} \\ \boldsymbol{F}^T & \boldsymbol{0} \end{pmatrix} \tag{3.10}$$

It should be noted that the $\boldsymbol{F}$ matrix is not necessarily symmetric, even if the waveguide is reciprocal. The reciprocity property is reflected in the full S-matrix, which, as can be seen from (3.10), is symmetric even if $\boldsymbol{F}$ is not.

In general, the matrix $\boldsymbol{F}$ is not diagonal; its nondiagonal elements represent mode coupling within the waveguide. A possible representation of a general, two-dimensional unitary matrix is

$$\boldsymbol{F} = e^{i\alpha} \begin{pmatrix} e^{i\phi}\cos\theta & e^{i(\pi-\psi)}\sin\theta \\ e^{i\psi}\sin\theta & e^{-i\phi}\cos\theta \end{pmatrix} \tag{3.11}$$

where α, ϕ, ψ, and θ are four real parameters. It can be seen that (3.7) is a special case of (3.11), with $\psi = \pi/2$. In Chapter 2 we considered the case of the polarization-preserving waveguide, which corresponds to $\theta = 0$, and

$$\alpha + \phi = \frac{2\pi L n_1 \nu}{c}, \qquad \alpha - \phi = \frac{2\pi L n_2 \nu}{c}$$

where L is the waveguide length, n_j is the effective refractive index for the jth mode, and ν is the optical field frequency.

The general S-matrix of a reciprocal two-port component can be written as

$$\boldsymbol{S} = \begin{pmatrix} \boldsymbol{R}_1 & \boldsymbol{T} \\ \boldsymbol{T}^T & \boldsymbol{R}_2 \end{pmatrix} \tag{3.12}$$

where the matrices $\boldsymbol{R}_1$ and $\boldsymbol{R}_2$ are symmetric. If the component preserves power, then the matrices $\boldsymbol{R}_1$, $\boldsymbol{R}_2$, and $\boldsymbol{T}$ satisfy the relations

$$\boldsymbol{R}_1\boldsymbol{R}_1^{\dagger} + \boldsymbol{T}\boldsymbol{T}^{\dagger} = \boldsymbol{I}$$

$$\boldsymbol{R}_2^{\dagger}\boldsymbol{R}_2 + \boldsymbol{T}^{\dagger}\boldsymbol{T} = \boldsymbol{I}$$

$$\boldsymbol{R}_1\boldsymbol{T}^{*} + \boldsymbol{T}\boldsymbol{R}_2^{\dagger} = \boldsymbol{0}$$

which follow from the condition that $\boldsymbol{S}$ is unitary.

3.3.3 Directional Couplers

Directional couplers are four-port devices that play an important role in optical networks. They are the optical analogs of the microwave hybrid junctions [2]. A scheme of a directional coupler is shown in Figure 3.3. An ideal directional coupler conserves power and does not reverse the power flow. In other words, all incoming power injected into ports 1 or 2 is routed to ports 3 or 4, and vice versa. The S-matrix of an ideal and reciprocal directional coupler has the form

$$S = \begin{pmatrix} \mathbf{0} & \mathbf{0} & \boldsymbol{J}_{13} & \boldsymbol{J}_{14} \\ \mathbf{0} & \mathbf{0} & \boldsymbol{J}_{23} & \boldsymbol{J}_{24} \\ \boldsymbol{J}_{13}^T & \boldsymbol{J}_{23}^T & \mathbf{0} & \mathbf{0} \\ \boldsymbol{J}_{14}^T & \boldsymbol{J}_{24}^T & \mathbf{0} & \mathbf{0} \end{pmatrix} \tag{3.13}$$

where the 4×4 matrix $\boldsymbol{M}$ defined by

$$\boldsymbol{M} = \begin{pmatrix} \boldsymbol{J}_{13} & \boldsymbol{J}_{14} \\ \boldsymbol{J}_{23} & \boldsymbol{J}_{24} \end{pmatrix}$$

is unitary. Thus, an ideal and reciprocal directional coupler is characterized by four Jones matrices, which together form a 4×4 unitary matrix.

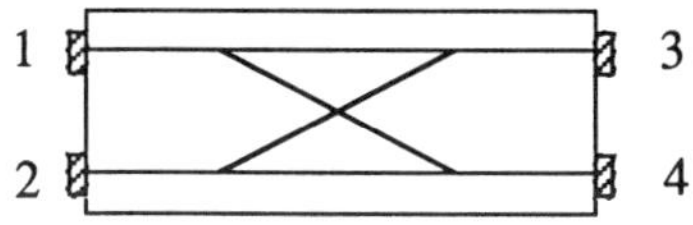

Figure 3.3 A scheme of a directional coupler with port assignments.

Quite often we deal with directional couplers that are also polarization-preserving. In such couplers, all four of the Jones matrices are diagonal, and like any polarization-preserving component, they can be logically separated into two strictly single-mode couplers. The S-submatrices $\boldsymbol{s}$ of the strictly single-mode directional couplers are of the order 4, and, assuming reciprocity, they have the form

$$\boldsymbol{s} = \begin{pmatrix} \mathbf{0} & \boldsymbol{K} \\ \boldsymbol{K}^T & \mathbf{0} \end{pmatrix} \tag{3.14}$$

where, for a power-preserving coupler, $\boldsymbol{K}$ is a unitary 2×2 matrix of the general form shown in (3.11). Formally, $\boldsymbol{s}$ resembles the S-matrix of an ideal waveguide (Eq. (3.10)).

3.4 PORT CHARACTERIZATION OF TIME-DEPENDENT COMPONENTS

Suppose that a Jones vector $\mathbf{A}_j(t)$ is incident upon port j of a linear, time-dependent N-port component. In general, $\mathbf{A}_j(t)$ will create a set of N Jones vectors $\mathbf{A}'_1$, ..., $\mathbf{A}'_N$, so that $\mathbf{A}'_k$ emerges from port k. The Jones vectors $\mathbf{A}'_k$ and $\mathbf{A}_j$ are related by

$$\mathbf{A}'_k(t_2) = \sum_j \int dt_1 \mathbf{J}_{kj}(t_2,t_1)\mathbf{A}_j(t_1) + \mathbf{C}_k(t_2) \tag{3.15}$$

where $\mathbf{J}_{kj}(t_2,t_1)$ is the transfer matrix between ports k and j. Using our "super vector" notation from Section 3.1, we can write (3.15) in the form

$$\tilde{\mathbf{A}}'(t_2) = \int dt_1 \mathbf{S}(t_2,t_1)\tilde{\mathbf{A}}(t_1) + \tilde{\mathbf{C}}(t_2) \tag{3.16}$$

where $\mathbf{S}(t_2,t_1)$ may be regarded in a certain sense as the time-domain S-matrix, and its elements $\mathbf{J}_{k,j}(t_2,t_1)$ as the time-domain scattering parameters. The corresponding frequency-domain S-matrix may be defined by the usual Fourier transform relation

$$S(\nu_2,\nu_1) = \int dt_1 dt_2 \exp[2\pi i(\nu_2 t_2 + \nu_1 t_1)]\mathbf{S}(t_2,t_1) \tag{3.17}$$

Note that S now depends on two frequencies, ν_1 and ν_2, in contrast with the single-frequency dependence of the corresponding S-matrix in the time-independent case.

The Fourier representation of cyclic Jones matrices was derived in Section 2.4. The same ideas may be applied to derive the Fourier representation of the S-matrix. Time-periodic components are described by cyclic S-matrices. A cyclic S-matrix $\mathbf{S}(t_2,t_1)$ may be represented as

$$\mathbf{S}(t_2,t_1) = \sum_n \exp(-2\pi i n f_0 t_2) \int d\nu s_n(\nu) \exp[-2\pi i \nu(t_2 - t_1)] \tag{3.18}$$

where the matrices $s_n(\nu)$ are the Fourier expansion coefficients of $\mathbf{S}$ and f_0 is the frequency of the component time variation. Inserting this representation in (3.16) and Fourier transforming, we obtain

$$\tilde{A}'(\nu) = \sum_n s_n(\nu - nf_0)\tilde{A}(\nu - nf_0) + \tilde{C}(\nu) \tag{3.19}$$

analogous to (2.24). The Fourier expansion coefficients of the time-domain S-matrix elements $\mathbf{J}_{kj}(t_2,t_1)$ are simply the matrix elements of $s_n(\nu)$.

3.5 SUMMARY

The S-matrix is a convenient parametrization of components. Although it is very familiar in the microwave context, its utilization in optics was not common until recently. The application of the S-matrix concept to optical components is problematic for two main reasons: (1) as yet there is no general experimental routine for the measurement of optical scattering parameters, and (2) these parameters are extremely sensitive to variations in external conditions, such as temperature and pressure. In spite of these practical difficulties, the S-matrix concept is very convenient for the *theoretical* analysis of optical networks.

The optical S-matrix differs from the familiar microwave S-matrix in that the matrix "elements" of the optical S-matrix are 2×2 matrices themselves, to account for the polarization. Thus, the optical scattering parameters are 2×2 matrices. In a proper context, these matrices may be regarded as transfer, or Jones, matrices.

The elements of the S-matrix are not independent. Two general physical principles impose certain relations between them. These principles are energy conservation and reciprocity. From the energy conservation principle it follows that the total power emerging from a passive component cannot exceed the total power that is fed into it. In practice, the output power is always smaller than the input power because of unavoidable losses. However, it is often convenient to consider an idealization, in which the losses are disregarded and the output power is exactly equal to the input power. The S-matrix of an ideally power-preserving component is unitary.

The reciprocity principle stems from certain symmetry properties of the Maxwell equations with respect to time reversal. This symmetry is destroyed in the presence of magnetic interactions. Another effect that destroys reciprocity is the Sagnac effect. In the absence of these effects, optical components can be regarded as reciprocal. The S-matrix of a reciprocal component is symmetric.

Optical components can be classified according to the number of their ports. We have considered in some detail certain one-, two- and four-port components. The class of three-port components was skipped, since such components are not very common. The only exception to this assertion is the Y coupler, which can be regarded as a four-port directional coupler with one port terminated.

Finally, we have briefly considered the S-matrix of time-periodic components, applying the same ideas that were used in the previous chapter to analyze the Jones matrix of such components.

References

[1] Carlin H., "The Scattering Matrix in Network Theory," *IRE Trans. Circuit Theory,* Vol. CT-3, 1956, pp. 89-97.

[2] Sander, K.F., *Microwave Components and Systems,* Wokingham: Addison-Wesley, 1987, pp. 142-145.

Further Reading

[1] Carlin H. and A. Giordano, *Network Theory: an Introduction to Reciprocal and Non-Reciprocal Circuits*, Englewood Cliffs: Prentice-Hall, 1964.

[2] Kim W.H. and H.E. Meadows, *Modern Network Analysis*, New York: John Wiley & Sons, 1971, pp. 237-265.

Chapter 4
The Signal Flow Graphs in Network Analysis

The aim of network analysis is to derive the network transfer function. This function allows us to determine the effect of the network on a given signal without any further reference to the network components or topology. In principle, network analysis should include a detailed treatment of the components and their interactions. However, by calling an assembly of components a network, it is implicitly assumed that the components can be treated as "black boxes," characterized by a set of parameters that are independent of the particular network implementation. In other words, it is assumed that the values of these parameters can be measured (or calculated) for each component independently, and these values can be used to derive the network transfer function. A possible choice for this set of parameters is the S-matrix described in Chapter 3.

Network analysis consists of two stages: the derivation of the network equations, and their solution. The unknowns in these equations are the guided fields at the network nodes, and the coefficients are the various scattering parameters of the network components. An attempt to derive the network equations directly from the physical scheme is sometimes confusing and prone to errors. It is much easier and safer to derive these equations from a scheme in which the relations between the unknown variables are displayed explicitly. The signal flow graph of the network [1] is most appropriate for this purpose.

Originally, signal flow graphs were introduced as a tool to solve the network equations. This motivation directed the main efforts in signal flow graph theory development [2]. This conception of signal flow graphs prevails in current textbooks dealing with network theory [3]. In these presentations, it is normally assumed that the signal flow graph is introduced only after the network equations are derived, and little attention is given to the question of how to associate a signal flow graph directly with the raw physical scheme of the network. Our motivation for using signal flow graphs is completely different: we regard them as a tool to facilitate the *derivation* of the network equations, rather than as a tool for their solution. Consequently, the method for the direct derivation of the signal flow graph from the network physical scheme will be considered here in detail, while most of the theoretical elaborations aimed at the graphical solution methods will be omitted. We will present, however, the elementary signal flow graph reduction rules [1]. These rules are useful for the solution of simple networks, and can be also used to simplify the graphs of complicated networks prior to the application of numerical procedures. We should emphasize, though, that the use of signal flow graphs is, in our view, a convenience rather than a necessity. The experienced analyst may wish to skip the signal flow graph stage and derive the network equations directly, especially in the analysis of simple networks.

It is always possible to write down the network equations in the time domain, in which the unknowns are time-dependent Jones vectors. We will see, however, that in the time-independent case it is advantageous to formulate the equations in the frequency domain. The frequency domain approach may be used in the time-periodic case as well, provided that the network algebra is modified accordingly. Even though the realization of the network equations, and consequently the method for their solution, differs in the various cases, the process of the derivation of the network equations, including the derivation of the network signal flow graph, is the same. Therefore, in the exposition we will use an abstract algebra so that later we will be able to apply the methods developed here in both the time and the frequency domains, and in the time-independent and the time-periodic cases.

4.1 GENERAL CONSIDERATIONS

At present, there are two generic classes of optical networks: those that use fiber-optic interconnects (fiber-optic networks) and those that use integrated waveguides (integrated-optics networks). There are, of course, also mixed networks, in which both types of waveguides are used. In fiber-optic networks, each component is supplied either with connectors or with fiber-optic leads, which can be spliced or connected to other optical fibers. In both cases, the ports are well defined physically. On the other hand, in integrated-optics networks the components reside on a common substrate, and there is no well-defined physical point that may correspond to a port. In this case, for network analysis purposes, we introduce for each component *virtual* ports, and are free to choose their physical location anywhere on the waveguide leads that are connected to it. Of course, the S-matrix of a component will depend on the particular choice of the physical location of its virtual ports.

This freedom introduces a certain ambiguity with regard to the question of whether the waveguides themselves are to be considered separate components. To illustrate this ambiguity, we show in Figure 4.1 a segment of an integrated-optics network containing two components *A* and *B* interconnected by a waveguide, with two valid choices of virtual port locations. In part (a), a single pair of ports *(a, b)* is introduced somewhere in the middle of the interconnecting waveguide. In this choice, the waveguide is not considered a component, and its sections between the component interaction regions and the corresponding ports are considered as "leads," i.e., parts of the corresponding components. In Figure 4.1(b), *two* pairs of ports are introduced: (a, w_1) and (w_2, b). Here the waveguide is considered as a separate component, with ports w_1 and w_2.

The "absorption" of the waveguides into the network components, as illustrated in Figure 4.1(a), simplifies the analysis, since it leads to a network scheme with fewer components. However, there are sometimes reasons for treating a particular waveguide or waveguides as separate components. This is the case, for instance, when investigating the effect of a waveguide parameter, most commonly its length, on the network transfer function. Another motivation for treating the waveguides as separate components is to introduce an approximation in which the scattering parameters of the components are frequency-independent. This approximation is very common, especially when highly coherent (narrow linewidth) optical sources are used. In this case, all the frequency dependence of the network transfer function is contained in the Jones matrices of the waveguides.

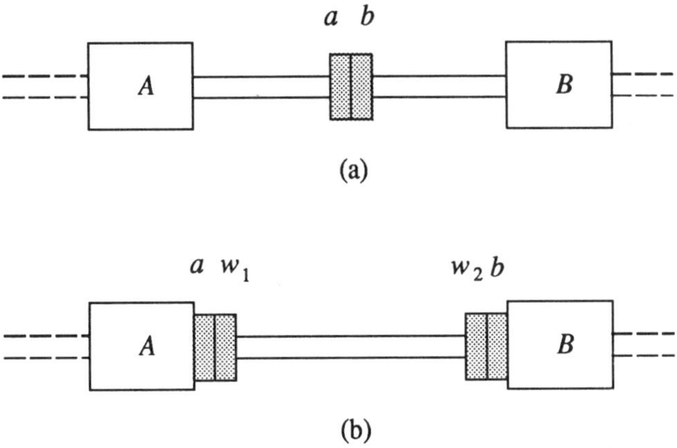

Figure 4.1 The two alternatives for the choice of virtual ports in the connection of two integrated-optics components: (a) the interconnecting waveguide divided into two sections, each considered a part of the corresponding component; (b) the interconnecting waveguide considered as a separate component.

Virtual ports may also be introduced instead of (or in addition to) the real ports in fiber-optic networks. This action may be even necessary, for instance, to account for the reflections or losses occurring at the splice of two fibers, and treat the splice itself as a two-port component. On the other hand, if the imperfections associated with the physical fiber connections are disregarded, then the fibers may be "absorbed" in the corresponding components, as shown in Figure 4.1(a).

The Jones and the S-matrices introduced in the two previous chapters were defined with respect to propagating guided fields only. However, in the interaction regions within the components, there are, in general, two other kinds of fields present: evanescent (nonpropagating) fields and radiative (unguided) fields. Rigorously speaking, to fully characterize the component, it is necessary to consider all three kinds of fields: guided, radiative, and evanescent. However, in contrast to the propagating guided fields, which contain two modes only, the radiative fields generally contain an infinite number of orthogonal modes; therefore, such a characterization would require an infinite S-matrix. Fortunately, it can be assumed that in most networks the components are allowed to interact by means of the propagating guided fields only. For the analysis of such networks, the characterization of the components with respect to the propagating guided fields alone is sufficient.

Radiative and evanescent field isolation of the components in an optical network is one of the primary goals of the network designer, because the evanescent and radiative field interactions may affect the network behavior in a manner that is hard to predict, and will usually degrade its performance. Evanescent field isolation can be effectively achieved by separating the interaction regions of any two connected components with waveguide leads that are long enough to allow the exponential decay of these fields to take place. It must be remembered, though, that the spatial decay distance of the various waveguide evanescent field modes varies with the optical frequency. Actually, at the frequency in which a particular evanescent field mode is transformed into a propagating mode, its decay distance naturally diverges to infinity. Thus, evanescent field isolation breaks down when the higher cutoff frequency is approached. In addition, in integrated-optics networks it may be necessary to separate the locations of the various components by a distance large enough to prevent the overlap of evanescent fields leaking into the substrate. By paying attention to all these precautions, evanescent field isolation may be achieved in both fiber-optic and integrated-optics networks. In fiber-optic networks, radiative field interactions between components can be effectively eliminated by protecting them with an opaque cover, thus allowing the radiation to enter and leave through the ports only. Such protection cannot be easily incorporated in integrated-optics networks, and the radiation field isolation of such networks is therefore less effective.

4.2 SIGNAL FLOW GRAPHS

4.2.1 The Signal Flow Graph as a Graphical Representation of Linear Algebraic Equations

Signal flow graphs may be considered a graphical representation of a linear set of equations. To illustrate the association between a signal flow graph and its corresponding set of equations, we consider as an example the following set of equations:

$$x_2 = t_1 x_1 + t_4 x_3$$

$$x_3 = t_2 x_1 + t_3 x_2$$

$$x_4 = t_5 x_3$$

Figure 4.2 shows the signal flow graph corresponding to this set of equations. The signal flow graph is composed of two entities: *nodes* and *directed branches*. To each node we assign a *variable*, and to each branch a *transmission*. In the graph shown in Figure 4.2, there are four nodes, and the variables that are associated with them are x_1, x_2, x_3, and x_4. In addition, there are five branches, and the transmissions assigned to them are t_1, t_2, t_3, t_4, and t_5. For each node, we divide the branches that are connected to it into two categories, according to their direction with respect to the chosen node: *incoming* and *outgoing*. For example, in Figure 4.2 the branch associated with t_2 is incoming with respect to the node associated with x_2, and outgoing with respect to the node associated with x_3. The values of the transmissions are usually assumed to be known.

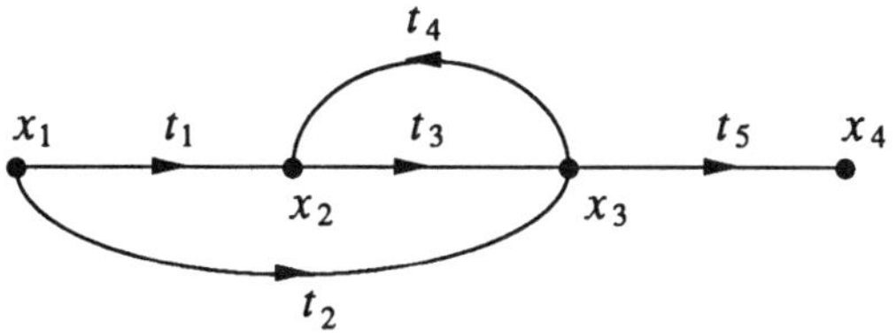

Figure 4.2 A signal flow graph illustrating a set of three linear equations.

The signal flow graph shown in Figure 4.2 has four nodes, but it is associated with only three equations. Thus, it is not possible to express the values of all the node variables in terms of the branch transmissions. This is a common situation in signal flow graphs: in order to solve the graph (i.e., to compute the values of the node variables), it is necessary to assume that not only the branch transmissions are known, but also some of the node variables. Having in mind the network application, this fact appears quite natural, since in order to compute the value of the signal(s) at the output port(s), it is generally necessary to specify not only the network parameters, but also the value of the signal(s) at the input port(s).

Returning to our example, we note that it is possible to express the values of the node variables x_2, x_3, and x_4 in terms of x_1 and the branch transmissions t_1, ..., t_5. In particular,

$$x_4 = \frac{t_1 t_3 + t_2}{1 - t_3 t_4} t_5 x_1$$

The signal flow graph corresponding to this equation is

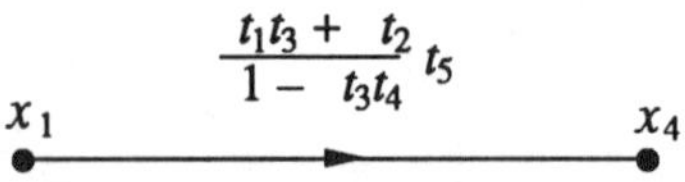

Figure 4.3 The solution of the graph shown in Figure 4.2.

If our aim was to express x_4 in terms of the branch transmissions and x_1, then this signal flow graph may be considered a solution of the signal flow graph shown in Figure 4.2.

4.2.2 The Association of Algebraic Equations With a Given Signal Flow Graph

After examining our simple example, we state the general rule for the derivation of the set of equations corresponding to a given signal flow graph:

The signal flow graph association rule: The set of equations corresponding to a signal flow graph is derived by taking the value of each node variable to be equal to the sum of all the incoming branch transmissions, each being multiplied by the node variable from which it originated.

Thus, in deriving the equation for a given node variable, all outgoing branches are ignored. The signal flow graph may be regarded as a graphical representation of the relations between the node variables. Each node acts as a signal source whose signal is being distributed to other nodes through outgoing branches. It is wrong, however, to regard the signal as being *divided* between the various outgoing branches; rather, imagine that the signal is being *duplicated* for each outgoing branch. One may wonder whether this procedure may violate energy conservation. This should not be a concern, however, because energy conservation will be guaranteed by a proper choice of branch transmissions values.

We divide the various nodes of a signal flow graph into three categories: *sources*, *sinks*, and *stars*. A source is a node that is connected to outgoing branches only. A sink is a node that is connected to incoming branches only. A star is a node that is connected to at least one incoming and one outgoing branch. In Figure 4.2, the node associated with x_1 is a source, the node associated with x_4 is a sink, and the rest are stars. Since a source node does not have any incoming branches by definition, no equation may be derived for a source variable. Therefore, the number of equations associated with a signal flow graph is equal to the number of its sink and star nodes, which is three in our example.

Ordinarily, there is interest only in the values of the sink node variables. These values can be immediately inferred if the graph contains only sink and source nodes.

Such a graph is called a *reduced* graph. Every signal flow graph can be reduced, provided, of course, that its corresponding set of equations is nonsingular. The signal flow graph shown in Figure 4.3 is the reduced form of the graph shown in Figure 4.2.

4.3 THE NETWORK ALGEBRA

In our example, the signal flow graph represented a set of linear algebraic equations. However, a signal flow graph can generally represent a set of any *operator* equations in which the branch transmissions are the operators and the node variables are the operands [4]. In order to be able to analyze the signal flow graph and its associated network, we have to define a *network algebra*, namely, the addition and the multiplication operations between any two given transmissions. To differentiate between the network algebra operations and other operations, we will denote the network algebra product of two transmissions u_1 and u_2 as $[u_1u_2]$ and their network algebra sum as $[u_1 + u_2]$.

In all cases of interest to us, the result ux of the operation of a transmission u on a node variable x is a node variable too. This fact allows us to define the product of two transmissions u_1 and u_2 as the equivalent of the two consecutive applications of u_1 and u_2 on x. In other words, u is defined as the *product* of u_1 and u_2, if for any node variable x

$$ux = [u_1u_2]\text{x} = u_1(u_2x) \tag{4.1}$$

The network product is noncommutative, so that in general $[u_1u_2] \neq [u_2u_1]$.

To define the network *addition* operation, we have to assume that the addition operation is defined for the node variables. In other words, if x_1 and x_2 are any two node variables, then there is an algorithm to compute $x_1 + x_2$, the result being a valid node variable. The transmission u is defined as the sum $[u_1 + u_2]$ of the two transmissions u_1 and u_2, if for any node variable x

$$ux = [u_1 + u_2]x = u_1x + u_2x \tag{4.2}$$

In all cases considered in this book, the addition of two node variables is commutative; therefore, the network algebra addition of two transmissions is also commutative.

In network applications, the transmissions depend on one or two continuous parameters α. These parameters are time variables in the time-domain representation, and frequency variables in the frequency-domain representation. To show this dependence explicitly in a product or a sum, we will use the notations $[u_1u_2](\alpha)$ and $[u_1 + u_2](\alpha)$, respectively.

Throughout this book, the network transmissions will represent different entities, depending on the case being considered. In the time domain, the network transmissions will represent Jones matrices that depend, in general, on two time arguments, and in the time-independent case on one time argument. In the frequency domain, a branch transmission will represent in the time-independent case a frequency-dependent Jones matrix, and in the time-periodic case a *set* of such matrices. A different network algebra will be defined for each case. The most fundamental case, from which all other algebra rules will be derived, is the case of the general time-domain representation. The general operation of a Jones matrix on a Jones vector in the time domain is defined by (2.10). The time-domain network product $[\boldsymbol{J}_1\boldsymbol{J}_2](t_2,t_1)$ of two Jones matrices $\boldsymbol{J}_1(t_2,t_1)$ and $\boldsymbol{J}_2(t_2,t_1)$ is derived from (4.1) and (2.10):

$$[\boldsymbol{J}_1\boldsymbol{J}_2](t_3,t_1) = \int dt_2 \boldsymbol{J}_1(t_3,t_2)\boldsymbol{J}_2(t_2,t_1) \tag{4.3}$$

The addition of two Jones vectors is defined as the familiar algebraic vector addition. Thus, if

$$\boldsymbol{A} = \begin{pmatrix} A_1 \\ \\ A_2 \end{pmatrix}, \quad \boldsymbol{B} = \begin{pmatrix} B_1 \\ \\ B_2 \end{pmatrix}$$

then

$$\boldsymbol{A} + \boldsymbol{B} = \begin{pmatrix} A_1 + B_1 \\ \\ A_2 + B_2 \end{pmatrix}$$

This allows us to define the network addition operation for the Jones matrices, which, from (4.2), is seen to be the familiar algebraic matrix addition:

$$[\mathbf{J}_1 + \mathbf{J}_2](t_2,t_1) = \mathbf{J}_1(t_2,t_1) + \mathbf{J}_2(t_2,t_1) \tag{4.4}$$

Equations (4.3) and (4.4) give the network algebra rules for the combination of two Jones matrices with an arbitrary time dependence. Later, we will consider in detail two special cases: the time-independent and time-periodic cases.

4.4 THE SIGNAL FLOW GRAPHS OF OPTICAL COMPONENTS

As we have seen, the S-matrix characterization of a component consists of a set of linear algebraic equations. This set of equations can be represented by a signal flow graph, which may be regarded as a graphical representation of the component. In this section we will derive the signal flow graphs of some commonly used optical components.

4.4.1 Single-Port Components

Let us start with the simple case of a single-port component. Such a component is described algebraically by (3.1). The signal flow graph corresponding to this equation is shown in Figure 4.4. The signal flow graph of a general single-port component consists, therefore, of three nodes and two branches. Two nodes are associated with the port, and one with the signal generator. One of the two nodes that represent the port is a source node, and the other a sink node. We will refer to these nodes as the *input* and the *output* nodes of the port, respectively. One of the two branches connects the signal generator to the output port with a unit transmission. The other branch connects the input and the output nodes, and its transmission is just the S-matrix of the component.

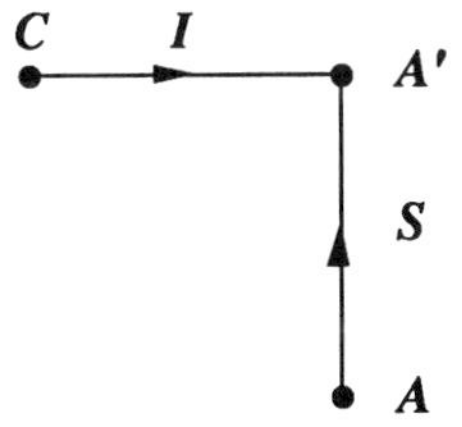

Figure 4.4 The signal flow graph of a general single-port component. This graph is also the full representation of an optical source.

The signal flow graph shown in Figure 4.4 may represent an *optical source*. In practice, however, it is often assumed that in the case of a source there are no reflections, i.e., that $\boldsymbol{S} = \boldsymbol{0}$. The equation describing such a source is simply

$$\boldsymbol{A}' = \boldsymbol{C} \tag{4.5}$$

In the signal flow graph corresponding to an optical source without reflections, the branch connecting the input and the output nodes vanishes, so that the input node becomes disconnected. Such a node can be ignored. Furthermore, the output node is connected now to the signal generator node only with a branch of transmission unity, and therefore the node variables corresponding to these two nodes become identical, as manifested in (4.5). Consequently, a source without reflections can be represented graphically by a single output node, with a corresponding node variable $\boldsymbol{C}$.

The optical detector is a passive single-port component. In such a component there is no signal generator node, so that it can be represented graphically by two nodes only. It is, however, convenient to attach a sink node to the input node of the detector signal flow graph. This operation has no effect on the relation between the input and the output node variables, but merely serves to emphasize the fact that the value of the detector input node has to be retained in the signal flow graph reduction process. Thus, the full signal flow graph of a detector has the form shown in Figure 4.5. As in the case of the optical source, it is often assumed that the S-matrix of a detector vanishes, i.e., that there are no reflections. Correspondingly, such a detector can be represented by a single *input* node, keeping in mind, though, that the variable associated with the hidden output node has the value of $\mathbf{0}$.

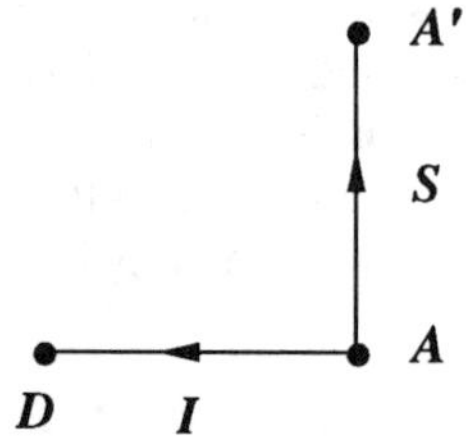

Figure 4.5 The full signal flow graph of a detector.

4.4.2 *N*-Port Components

Having examined the simple case of one-port components, it should be clear how to proceed to the general case. The set of equations that make up the S-matrix characterization of an N-port component is given in (3.3). This set of equations contains, altogether, $3N$ vector variables, $2N$ corresponding to the Jones vectors $\boldsymbol{A}_1$, $\boldsymbol{A}_2$, ..., $\boldsymbol{A}_N$, and $\boldsymbol{A}'_1$, $\boldsymbol{A}'_2$, ..., $\boldsymbol{A}'_N$, and N corresponding to the source variables $\boldsymbol{C}_1$, $\boldsymbol{C}_2$, ..., $\boldsymbol{C}_N$. Accordingly, in the corresponding signal flow graph, each port is represented by two nodes, one being associated with the incoming Jones vector, and the other with the outgoing Jones vector. These are the input and the output port nodes, respectively. There are N^2 branches that connect the input nodes to the output nodes. The transmissions of these branches are the corresponding S-matrix elements of the component. The source variables $\boldsymbol{C}_1$, $\boldsymbol{C}_2$, ..., $\boldsymbol{C}_N$ are connected to the output nodes with branches of transmission unity.

In Figure 4.6 we show a portion of the signal flow graph of a passive N-port component containing the nodes corresponding to the ports j and k. For clarity, we have omitted the node variables and the branch transmission symbols from the graph.

Instead, we have denoted the nodes with their port name, which will usually be an integer or a letter representing it. To distinguish between the names of the input and the output nodes, we put a prime on the name of the output node. This convention will be frequently used below. Note also that there are no stars in the graph of Figure 4.6, so that the signal flow graph of an isolated component is always reduced. (This assertion is true for active components as well, since the source nodes are connected only to the sink nodes of the ports.)

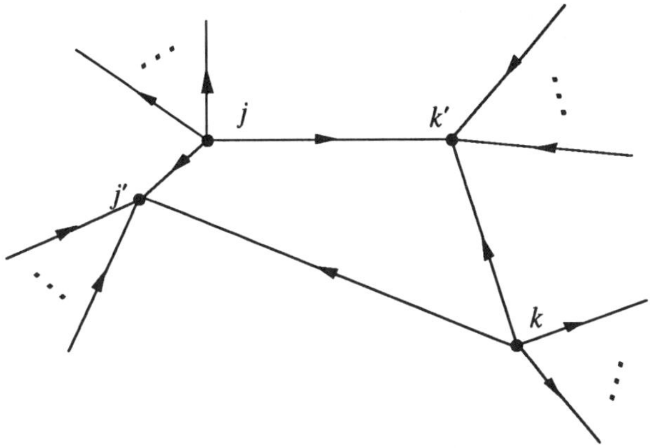

Figure 4.6 A portion of the signal flow graph of a general component showing two ports, *j* and *k*.

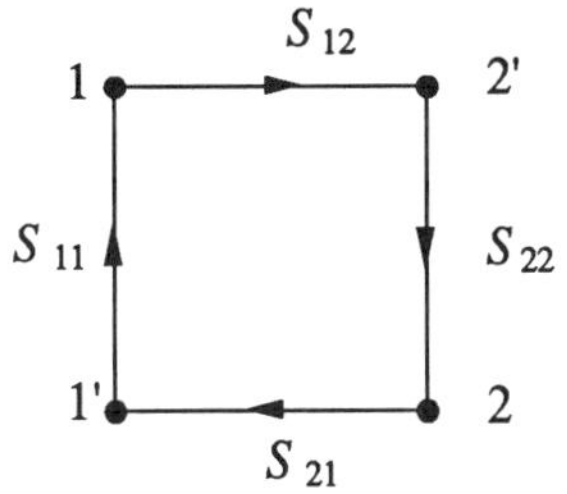

Figure 4.7 The signal flow graph of a passive, two-port component.

4.4.3 Two-Port Component

After the exposition of the general case, we return to consider some common special cases. The general signal flow graph of a passive, two-port component has the form shown in Figure 4.7. According to the general rule, we associated two nodes with each port, ending up with four nodes altogether. Every input node is connected with every output node by a directed branch. The branch transmissions are the component S-matrix elements, and may be considered Jones matrices. We remind the reader that if the component is reciprocal, then the matrices $\boldsymbol{S}_{11}$ and $\boldsymbol{S}_{22}$ are symmetric, and

$$\boldsymbol{S}_{12} = \boldsymbol{S}_{21}{}^{T}$$

The matrices $\boldsymbol{S}_{11}$ and $\boldsymbol{S}_{22}$ represent reflections. These matrices vanish in the case of an ideal waveguide, since it is assumed that an ideal waveguide has no reflections. Consequently, the signal flow graph of an ideal waveguide has the form shown in Figure 4.8.

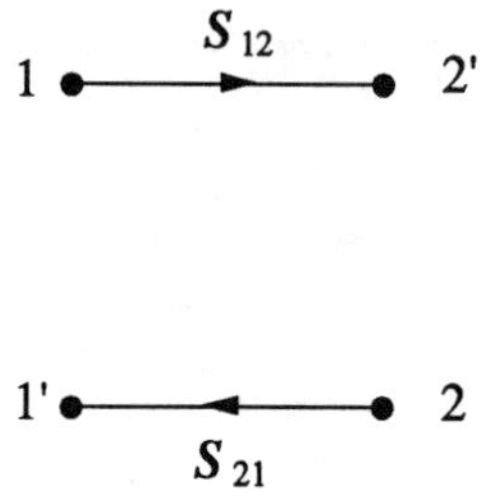

Figure 4.8 The signal flow graph of an ideal waveguide.

4.4.4 Directional Couplers

We turn now to the case of the directional coupler. Equation (3.13) shows the S-matrix of this device, and the corresponding signal flow graph is shown in Figure 4.9.

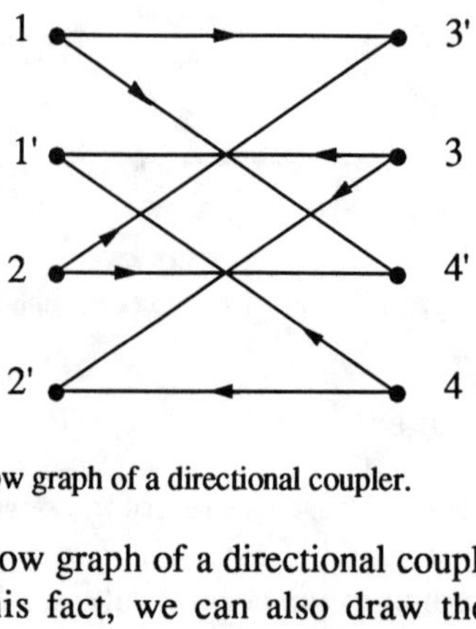

Figure 4.9 The signal flow graph of a directional coupler.

It is seen that the signal flow graph of a directional coupler consists of two disjoint graphs. To emphasize this fact, we can also draw the signal flow graph of a directional coupler in the form shown in Figure 4.10. Thus, the signal flow graph of a directional coupler is formally equivalent to the signal flow graphs of two two-port components.

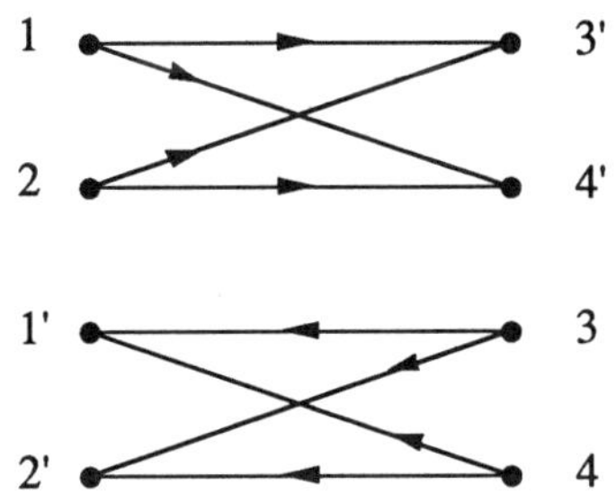

Figure 4.10 The disjoint representation of the signal flow graph of a directional coupler.

4.5 THE DERIVATION OF SIGNAL FLOW GRAPHS FOR OPTICAL NETWORKS

In order to derive the signal flow graph of the network, we have to describe the method for the connection of the signal flow graphs of the individual components. To formulate the connection process analytically, let us consider two components characterized by

$$\tilde{\boldsymbol{A}}' = \boldsymbol{u}\tilde{\boldsymbol{A}} + \tilde{\boldsymbol{a}} \tag{4.6}$$

and

$$\tilde{\boldsymbol{B}}' = \boldsymbol{v}\tilde{\boldsymbol{B}} + \tilde{\boldsymbol{b}} \tag{4.7}$$

The S-matrices $\boldsymbol{u}$ and $\boldsymbol{v}$ can generally have different dimensions. Suppose now that port j of the first component is connected to port k of the second one. As a result, the Jones vector $\tilde{\boldsymbol{B}}_k$ incident on the kth port of the second component becomes the Jones vector $\tilde{\boldsymbol{A}}'_j$ emerging from the jth port of the first component. Similarly, the Jones vector $\tilde{\boldsymbol{A}}_j$ incident on the jth port of the first component becomes the Jones vector $\tilde{\boldsymbol{B}}'_k$ emerging from the kth port of the second one. Thus, the description of the connected components amounts to Equations (4.6) and (4.7) augmented by

$$\tilde{\boldsymbol{B}}_k = \tilde{\boldsymbol{A}}'_j \tag{4.8}$$

and

$$\tilde{\boldsymbol{B}}'_k = \tilde{\boldsymbol{A}}_j \tag{4.9}$$

In the signal flow graph picture, the connection is accomplished by connecting node j' to node k with a branch of transmission unity directed from j' to k, and similarly connecting node k' to node j. An elementary application of the signal flow graph association rule shows that the set of equations corresponding to the graph combined in this manner is indeed composed of Equations (4.6) to (4.9), which demonstrates the validity of this connection procedure.

The graphical connection process described above, although perfectly valid, is excessively detailed. Since (4.8) and (4.9) establish an identity between the variables associated with nodes k and j', and between the variables associated with the nodes k' and j, we may as well *merge* these nodes respectively instead of connecting them with branches of unity transmission as suggested above. This action will result in the annihilation of two nodes, but will not inflict any harm, because the variables associated with them are redundant. This modified connection procedure is preferred since it yields a more compact result. We can state now the signal flow graph connection rule:

The signal flow graph connection rule: When two ports a and b are connected, the output node of a is merged with the input node of b, and the input node of a is merged with the output node of b.

This connection procedure is illustrated in Figure 4.11. Note that two connected ports are represented in the signal flow graph by two nodes only. The names of these nodes can be chosen arbitrarily; however, it is usually convenient to retain for these nodes the names of the original nodes of the more complex component (the component with the larger number of nodes, for instance). We draw attention to the fact that in the connection procedure the branches are preserved; the number of branches in the connected graph equals the total number of branches of the graphs before the connection. The connection procedure usually creates star nodes.

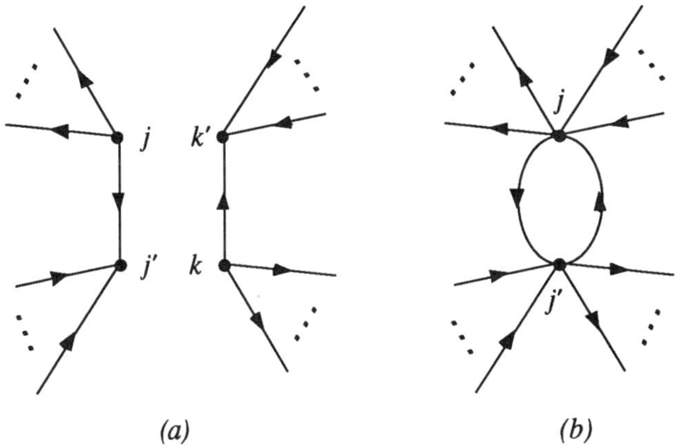

Figure 4.11 The connection process of two ports j and k: (a) prior to connection; (b) after the connection.

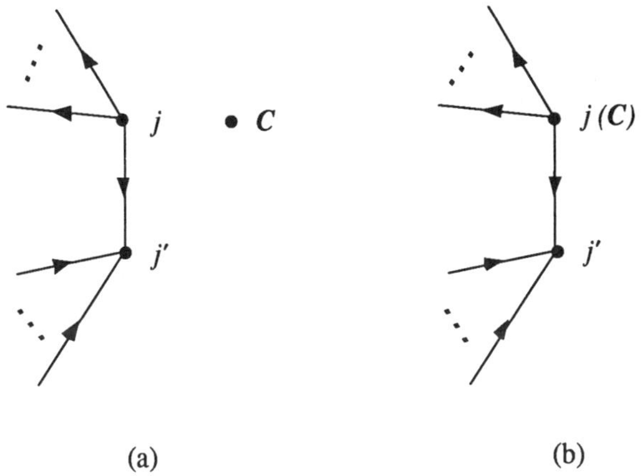

Figure 4.12 The simplified connection of an optical source to a port j: (a) prior to connection; (b) after the connection.

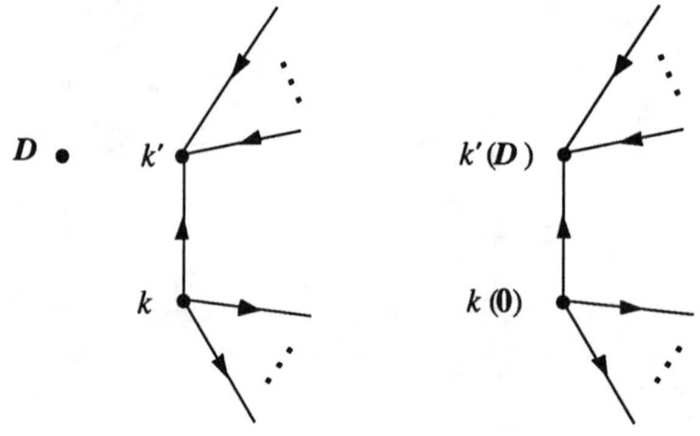

Figure 4.13 The simplified connection of an optical detector to a port k: (a) prior to connection; (b) after the connection.

The signal flow graph connection rule cannot be applied to the connection of optical sources or detectors if the simplified single-node representation of these components is used. The corresponding connection rules are deduced from the general rule and the full signal flow graphs of these components (Figures 4.3 and 4.4). The node representing the optical source is merged with the input node of the corresponding port, as shown in Figure 4.12. A single merge is also performed in the connection of a component to a single-node detector. However, in this case, since it is assumed that there are no reflections, the node variable associated with the input node of the corresponding port assumes the value of $\mathbf{0}$, as shown in Figure 4.13, thus becoming a null source. Such a node can be removed from the signal flow graph together with all its outgoing branches without otherwise affecting the graph. Note that in Figures 4.12 and 4.13 we have added in parentheses the names or the values of the corresponding variables to the nodes that were merged with the source and detector nodes. We will make further use of this notation in the examples that follow.

4.6 GRAPHICAL REDUCTION RULES FOR THE OPTICAL SIGNAL FLOW GRAPHS

Generally, the step following the derivation of the network signal flow graph is the formulation of the network equations and their solution. In certain cases, however, it may be convenient to derive the solution directly from the signal flow graph, by using certain graphical reduction rules. These rules may also be useful for simplifying the graph before proceeding to the equation formulation stage.

4.6.1 Branches in Series

The first rule that we consider concerns the reduction of two branches appearing in series. Suppose that a signal flow graph contains the following segment:

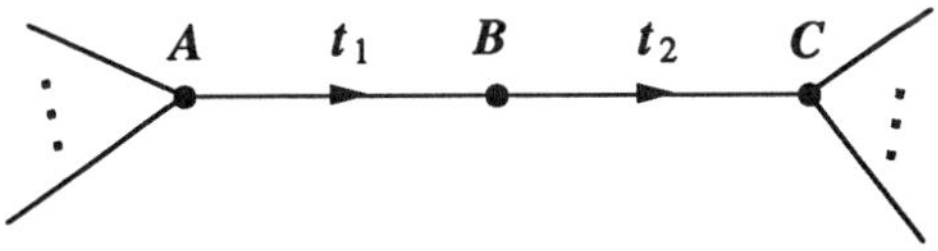

Figure 4.14 Two branches connected in series.

The variables A, B, and C satisfy the following equations:

$$B = t_1 A$$

$$C = t_2 B + Q$$

where Q stands for the possible contribution of incoming branches other then the one associated with t_2. Substituting $t_1 A$ for B in the second equation, we get

$$C = [t_2 t_1] A + Q$$

The signal flow graph corresponding to this equation is simply

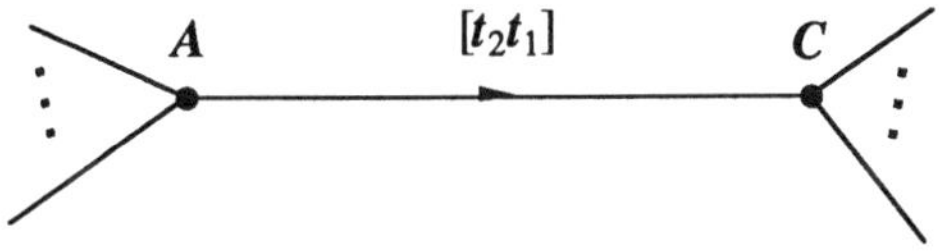

Figure 4.15 The equivalent of two branches in series.

Note that we have implicitly assumed that the two branches shown in Figure 4.14 are also the only branches that are connected to the node associated with B in the original graph. This simple manipulation allows us to formulate the first graphical reduction rule:

First reduction rule (branches in series): A signal flow graph segment consisting of node k and two branches (j,k) and (k,n) can be replaced by a single branch (j,n) with a transmission of $[t_{k,n} t_{j,k}]$ This rule may be applied only if the branches (j,k) and (k,n) are the only branches connected to the node k in the original graph.

In the formulation of the first reduction rule we have introduced a new notation: we have referred to a branch connecting nodes j and k and directed from j to k as (j,k), and to its transmission as $t_{j,k}$. We note that this name assignment is not unique, since, in general, there may be more than one branch directed from j to k. In order to distinguish between different branches connecting the same two nodes in the same direction we will use an extra superscript, such as $(j,k)^{(1)}$ and $(j,k)^{(2)}$.

4.6.2 Branches in Parallel

The second rule concerns two branches connected in parallel. Consider the following signal flow graph segment:

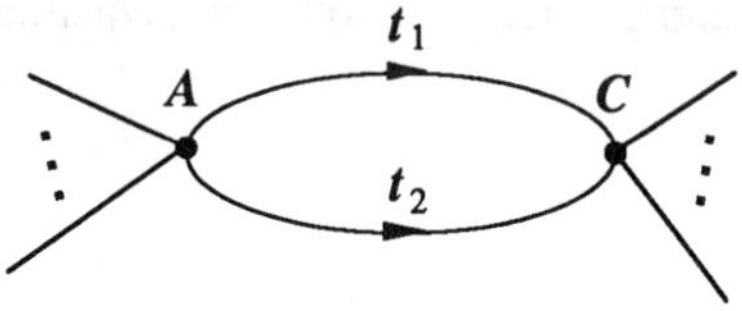

Figure 4.16 Two branches connected in parallel.

We assume that in the original graph the only branches that connect the nodes associated with the variables A and C are those shown in Figure 4.16. Under this assumption, these variables satisfy the following equation:

$$C = [t_1 + t_2]A + Q$$

with Q having the same meaning as above. Evidently, the two branches associated with t_1 and t_2 can be replaced by a single one having the transmission $[t_1 + t_2]$:

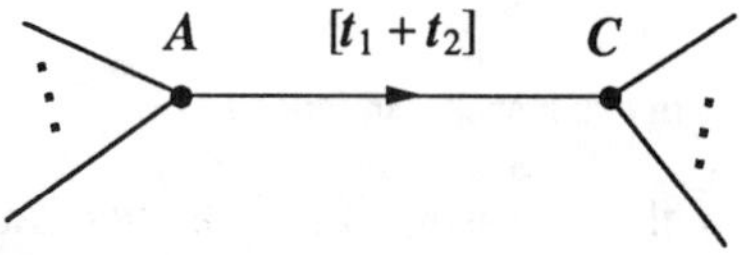

Figure 4.17 The equivalent of two branches connected in parallel.

This leads us to the formulation of the second reduction rule:

Second reduction rule (branches in parallel): A signal flow graph segment consisting of two branches $(j,k)^{(1)}$ and $(j,k)^{(2)}$ can be replaced by a single branch (j,k) with a transmission equal to the sum of the transmissions of the two original branches.

4.6.3 Elimination of a Feedback Branch

We turn now to a slightly more complicated case, which is referred to in the literature as a *feedback branch* [1]. Figure 4.18 portrays a signal flow graph segment containing such a branch:

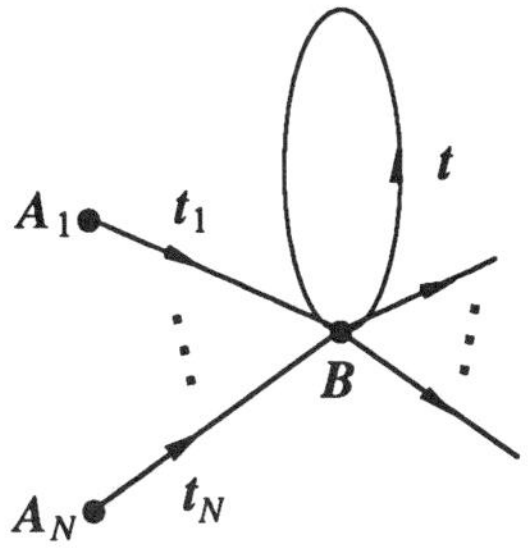

Figure 4.18 A signal flow graph segment containing a feedback branch.

The variables $\boldsymbol{A}$ and $\boldsymbol{B}$ are related as follows:

$$\boldsymbol{B} = \sum_k \boldsymbol{t}_k \boldsymbol{A}_k + \boldsymbol{t}\boldsymbol{B}$$

Formally, we may write the following explicit expression for $\boldsymbol{B}$:

$$\boldsymbol{B} = \sum_k [(\boldsymbol{I} - \boldsymbol{t})^{-1} \boldsymbol{t}_k] \boldsymbol{A}_k$$

The signal flow graph corresponding to this equation is shown in Figure 4.19. The feedback branch elimination rule can thus be formulated as follows:

Third reduction rule (feedback branch elimination): Let p be a node connected altogether to N incoming branches with transmissions $\boldsymbol{t}_1, \boldsymbol{t}_2, \ldots, \boldsymbol{t}_N$ and a feedback branch with a transmission $\boldsymbol{t}$. The feedback branch can be eliminated if all the transmissions of the incoming branches are modified so that $\boldsymbol{t}_k$ is replaced by $[(\boldsymbol{I} - \boldsymbol{t})^{-1}\boldsymbol{t}_k]$. *Remark:* The node p in the original graph may be connected to outgoing branches as well. These branches remain intact during the feedback branch elimination procedure.

The elimination of a feedback branch requires the inversion of a transmission. In the time-independent case, we will see that such an inversion can be reduced simply to the algebraic inversion of a 2×2 matrix by using the frequency-domain representation. However, in general time-dependent networks such an inversion may be rather difficult to accomplish.

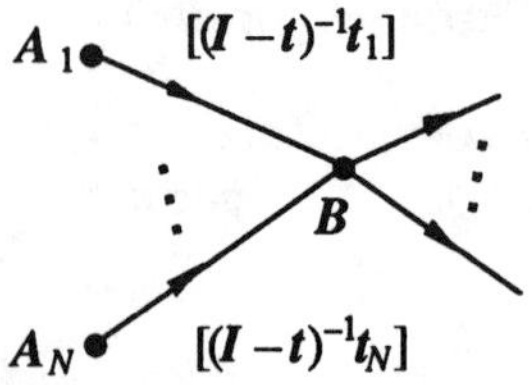

Figure 4.19 The equivalent of Figure 4.18.

4.6.4 Expansion of Stars

It often occurs that, after the elimination of all parallel, series, and feedback branches, the resulting graph still contains stars. In order to reduce such graphs further, we need an additional reduction rule.

A general star is shown in Figure 4.20. The nodes corresponding to the variables $\boldsymbol{A}_j$ and $\boldsymbol{C}_j$ may be connected to other nodes in addition to the one corresponding to $\boldsymbol{B}$.

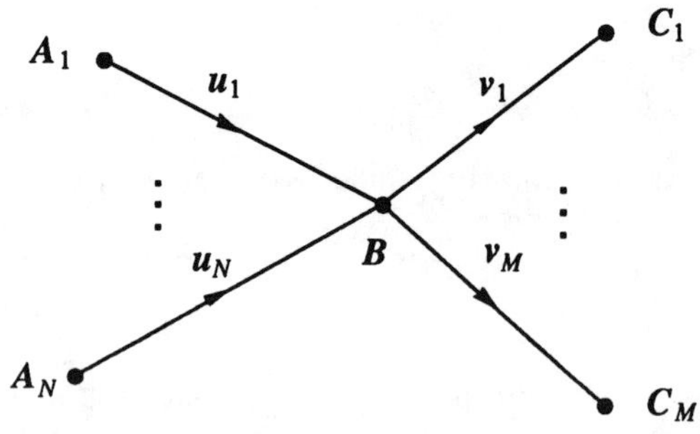

Figure 4.20 A signal flow graph segment containing a star.

The set of equations corresponding to the variables $\boldsymbol{B}$ and $\boldsymbol{C}$ is

$$\boldsymbol{B} = \sum_k u_k \boldsymbol{A}_k$$

$$\boldsymbol{C}_j = v_j\boldsymbol{B} + \boldsymbol{Q}_j = \sum_k [v_j u_k]\boldsymbol{A}_k + \boldsymbol{Q}_j, \quad j = 1, ..., M$$

where again $\boldsymbol{Q}_j$ stands for the possible contribution to $\boldsymbol{C}_j$ coming from any branch different from the one associated with v_j. The last set of equations is implied by the signal flow graph shown in Figure 4.21. For the sake of clarity, we have shown in Figure 4.21 only one node representing the group $\boldsymbol{C}_1, \boldsymbol{C}_2, ..., \boldsymbol{C}_M$.

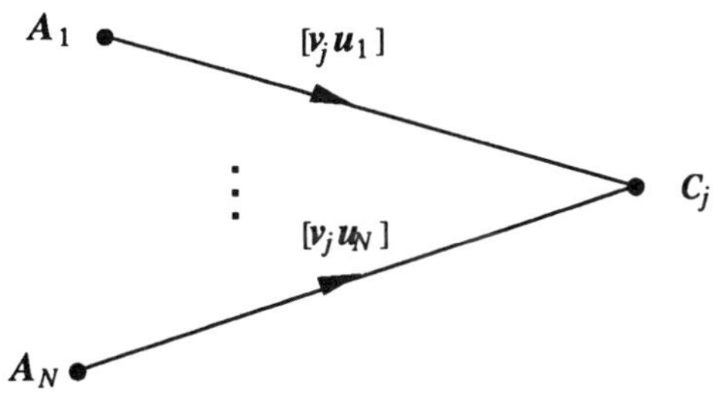

Figure 4.21 A section of the expanded version of the graph shown in Figure 4.20.

These considerations allow us to formulate the fourth and final signal flow graph reduction rule:

Fourth reduction rule (star elimination): Let Σ be a signal flow graph segment consisting of a node p, N branches (n_k,p) with transmissions $\boldsymbol{u}_k$, and M branches (p,m_j) with transmissions $\boldsymbol{v}_j$. The segment Σ may be replaced by a set of $N \times M$ branches (n_k,m_j) with transmissions $[\boldsymbol{v}_j\boldsymbol{u}_k]$.

It can be seen that the first reduction rule is a special case of the fourth. We have nevertheless stated the first rule on its own, since in practice the occurrence of branches that are connected in series is very common. We note that the fourth reduction rule may be applied even if for certain j and k, $n_k = m_j$. To illustrate this fact, let us consider the signal flow graph shown in Figure 4.22.

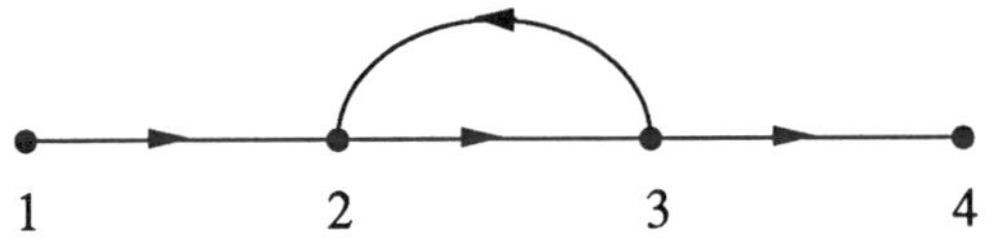

Figure 4.22 A signal flow graph containing a star, the elimination of which creates a feedback branch.

Let us apply the star elimination rule to node 3. This creates a feedback branch connected to node 2, as shown in Figure 4.23(a). Applying the third reduction rule to eliminate it, we obtain the final form shown in Figure 4.23(b). An equally valid procedure would be to start the reduction of the signal flow graph shown in Figure 4.22 by eliminating node 2 instead of node 3. This would lead to the graphs shown in Figure 4.24.

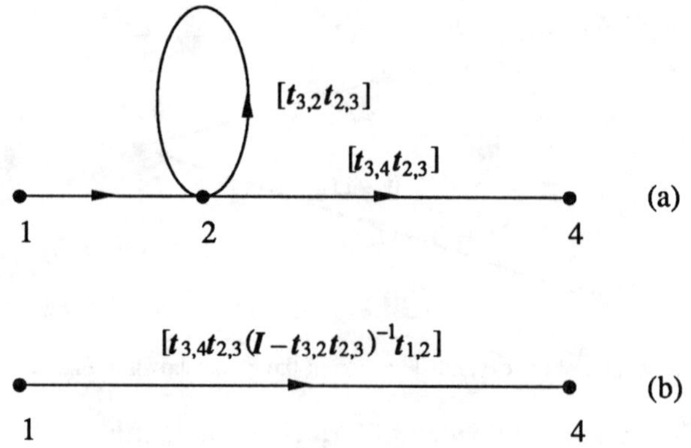

Figure 4.23 The reduction of the signal flow graph shown in Figure 4.22 by the elimination of node 3: (a) the signal flow graph after the elimination of node 3; (b) the signal flow graph after the elimination of the feedback branch.

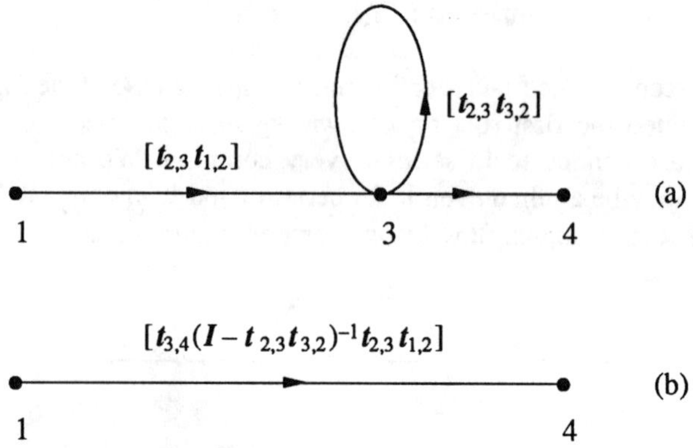

Figure 4.24 The reduction of the signal flow graph shown in Figure 4.22 by the elimination of node 2: (a) the signal flow graph after the elimination of node 2; (b) the signal flow graph after the elimination of the feedback branch.

At first sight it may seem that these two procedures yield different results, since the transmissions $t_{2,3}$ and $t_{3,2}$ do not necessarily commute. However, it is an elementary exercise in algebra to show that

$$[(I - t_{2,3}t_{3,2})^{-1}t_{2,3}] = [t_{2,3}(I - t_{3,2}t_{2,3})^{-1}]$$

for any two transmissions $t_{2,3}$ and $t_{3,2}$.

4.7 SUMMARY

The signal flow graph is a general tool for the analysis of networks. Originally, one of the purposes of signal flow graphs was the derivation of the network transfer function by the successive application of certain topological theorems. Nowadays, except perhaps in the simplest cases, it is more practical to use numerical methods for this purpose. However, other features of signal flow graphs are still useful.

Algebraically, the derivation of the transfer function of a linear network amounts to the solution of a set of algebraic equations. The network analysis usually starts with the physical scheme. A direct derivation of the network equations from the physical scheme is often confusing, especially in cases in which there are ports through which the field propagates in both directions. With the signal flow graph we can divide the process of the network equations derivation into two independent stages: the derivation of the signal flow graph and the derivation of the network equations. Each of these stages is much simpler to accomplish than their combination, i.e., the direct derivation of the network equations from the physical scheme. Although the derivation of the signal flow graph requires an extra effort, it may sometimes prevent the occurrence of mistakes.

The signal flow graph consists of two kinds of objects: nodes and directed branches. With each node there is an associated Jones vector, and with each branch a Jones matrix. Every optical component has a signal flow graph representation. With each port there are two associated nodes, one representing an incoming field and the other an outgoing field. Each input node is connected to each output node by a directed branch.

The derivation of the network signal flow graph consists of two stages. In the first stage, we replace every individual component of the network by its signal flow graph. Next, these signal flow graphs are connected by merging the corresponding input and output nodes. The network equations are derived by applying the signal flow graph correspondence rule. Quite often the signal flow graph can be simplified by applying certain elementary graphical reduction rules. There are four useful graphical reduction rules: combination of branches connected in series, combination of branches connected in parallel, feedback loop elimination, and star expansion. In a typical application, the graph is first simplified using the graphical reduction rules, and then the network equations are derived and solved numerically.

The use of signal flow graphs often inspires physical intuition and helps in the understanding of the network function. The combination of graphical and numerical methods is the most effective approach for both gaining physical insight and deriving the solution for the network problem.

References

[1] Mason, S.J., "Feedback Theory – Some Properties of Signal Flow Graphs," *Proc. IRE.*, Vol. 41, 1953, pp. 1144-1156.

[2] Mason, S.J., "Feedback Theory – Further Properties of Signal Flow Graphs," *Proc. IRE*, Vol. 44, 1956, pp. 920-926.

[3] Kim, W.H. and H.E. Meadows, *Modern Network Analysis*, New York: John Wiley, 1971, pp. 324-346.

[4] Robichaud, L.P.A., M. Boisvert, and J. Robert, *Signal Flow Graphs and Applications*, Englewood Cliffs: Prentice-Hall, 1962, pp. 182-186.

Further Reading

[1] Chow, Y. and E. Cassignol, *Linear Signal Flow Graphs and Applications*, New York: John Wiley, 1962.

Chapter 5
The Analysis of Time-Independent Networks

The most straightforward application of the methods that were developed in the previous chapters is in the analysis of time-independent networks. Using the frequency-domain representation, the network algebra is reduced in this case to the familiar matrix algebra, and the network equations become simply a set of linear algebraic equations. In order to illustrate the methodology, we present in detail the analysis of four simple, yet common, networks. Many readers may feel that in these simple cases the final result could be inferred immediately, without invoking the elaborate process of the signal flow graph derivation and its reduction. This is undoubtedly true. The usefulness of the methods that we advocate here becomes apparent in the analysis of more complex cases, where intuitive or adhoc approaches are likely to fail.

Three of the networks that we consider here are named after well-known unguided optical systems: the Fabry-Perot, the Mach-Zehnder, and the Michelson interferometers. These networks may be regarded as the guided implementation of the corresponding unguided systems. In some respects, the unguided interferometric systems behave like their guided versions. This is due to the fact that in many interferometric setups the system response is determined by the fundamental mode. As far as only the fundamental mode is concerned, there is no formal difference between the guided and unguided systems.

In the examples that we consider below, we will make frequent use of optical sources, detectors, matched terminations, waveguides, polarizers, and directional couplers. It is therefore convenient to introduce special symbols that will represent these components in the physical schemes. These symbols are shown in Figure 5.1. For the sake of clarity, we have indicated the ports explicitly. As discussed earlier in Section 4.1, these ports may correspond to real physical entities, or be virtual.

5.1 THE NETWORK ALGEBRA RULES FOR THE TIME-INDEPENDENT CASE

In the time-independent case we can define two different rules for multiplication corresponding to the time- and frequency-domain representations. In the time domain, the transmissions are 2×2 complex matrix functions of time. We remind the reader that although fundamentally the Jones matrix is a function of two time arguments, in the time-independent case it is a function of a single argument only. This single argument is the difference between the original time arguments. From (4.3) it follows that the network algebra product $[\mathbf{J}_2\mathbf{J}_1]$ of the transmissions $\mathbf{J}_1(t)$ and $\mathbf{J}_2(t)$ is the convolution integral

$$[\mathbf{J}_2\mathbf{J}_1](t) = \int d\tau \mathbf{J}_2(t-\tau)\mathbf{J}_1(\tau) \tag{5.1}$$

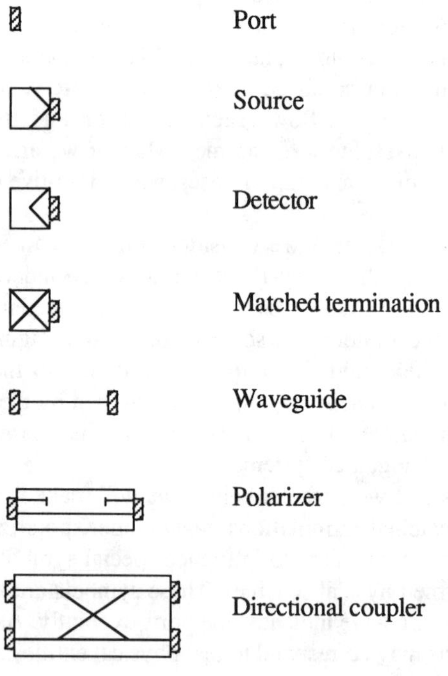

Figure 5.1 Commonly occurring symbols in physical schemes of optical networks.

In the frequency domain, the transmissions are 2×2 complex matrix functions $J(\nu)$ of the frequency ν, where $J(\nu)$ is simply the Fourier transform of $\mathbf{J}(t)$ (2.11).

Since (5.1) is a convolution integral, it is clear that the network algebra product [$\mathbf{J}_2\mathbf{J}_1$] of two transmissions $J_1(\nu)$ and $J_2(\nu)$ is simply the familiar algebraic matrix product of two matrices:

$$[\mathbf{J}_2\mathbf{J}_1](\nu) = J_2(\nu)J_1(\nu) \tag{5.2}$$

The network algebra addition of two transmissions is the same in both domains, and amounts to the familiar matrix addition of two matrices.

It is quite obvious that the network algebra product is much simpler in the frequency domain than in the time domain. It is therefore significantly advantageous to carry out the analysis of time-independent networks in the frequency domain. Accordingly, we will carry out the analyses throughout this chapter in the frequency domain. It should be noted, however, that the network algebra product in the time domain of two components with an instantaneous time response (2.12) reduces to a simple matrix product as well. Therefore, there is no advantage in using the frequency-domain representation for the analysis of networks consisting of instantaneous components only.

5.2 THE GUIDED-WAVE FABRY-PEROT INTERFEROMETER

The first example that we consider is the guided-wave version of a Fabry-Perot interferometer. The physical scheme of this system is shown in Figure 5.2(a). The two port components M_1 and M_2 play the role of the partially reflecting mirrors in the conventional Fabry-Perot interferometer. There are, altogether, eight ports in the physical scheme, and they are marked by the integers 1, 2, ..., 8. Figure 5.2(b) shows the individual signal flow graphs of the various components prior to connection. We have used the single-node representation for the optical source and the detector for simplicity. Application of the connection rule results in the signal flow graph shown in Figure 5.2(c). Two nodes in this signal flow graph are irrelevant, and can be discarded: node 7 is a null source, and node 2' is a sink whose node variable represents the Jones vector incident on the source. As we have mentioned already, a null source can always be removed without otherwise affecting the graph. Likewise, a sink may be removed, since its presence does not affect the value of any other network node. Note that when a node is removed from a signal flow graph, all the branches connected to it are removed as well. After removing nodes 2' and 7 we get the signal flow graph shown in Figure 5.2(d).

Let us turn now to the task of solving the Fabry-Perot signal flow graph. Applying the first reduction rule to the segment consisting of the nodes 6, 6', 3, 3' and the branches (6,6'), (6',3), and (3,3'), we get the graph shown in Figure 5.3(a). The graph in Figure 5.3(b) is derived by applying the fourth reduction rule for the elimination of node 6. This operation creates a feedback branch attached to node 3'.

This feedback branch can be eliminated with the help of the third reduction rule, which leads to the reduced graph shown in Figure 5.3(c).

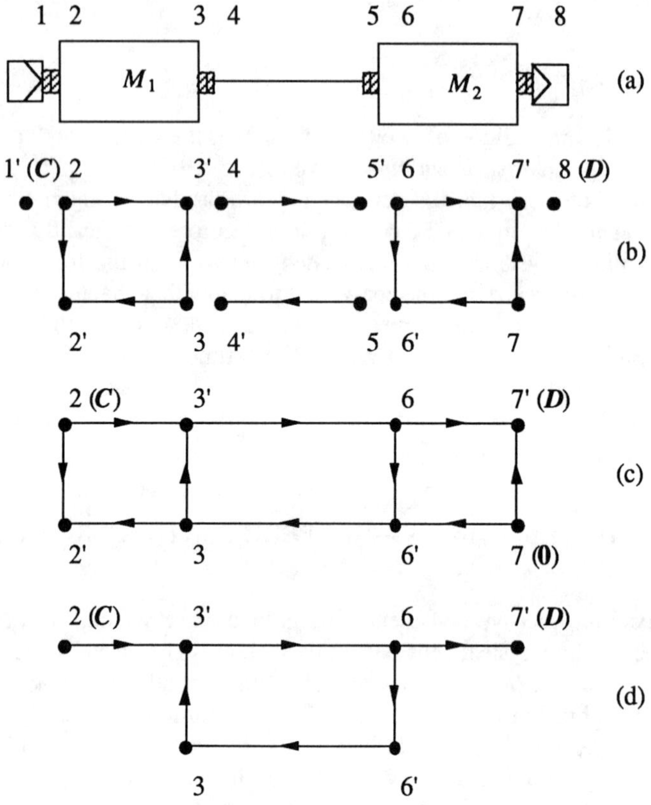

Figure 5.2 The derivation of the signal flow graph corresponding to a guided-wave Fabry-Perot interferometer: (a) the physical scheme; (b) the signal flow graphs of the various components before the connection; (c) the signal flow graph after the connection; (d) the final signal flow graph after the removal of the irrelevant nodes and branches.

Let us assume for simplicity that the components M_1 and M_2 are identical and reciprocal, and that their S matrix S_m has the form

$$S_m = \begin{pmatrix} r & t \\ t^T & r \end{pmatrix}$$

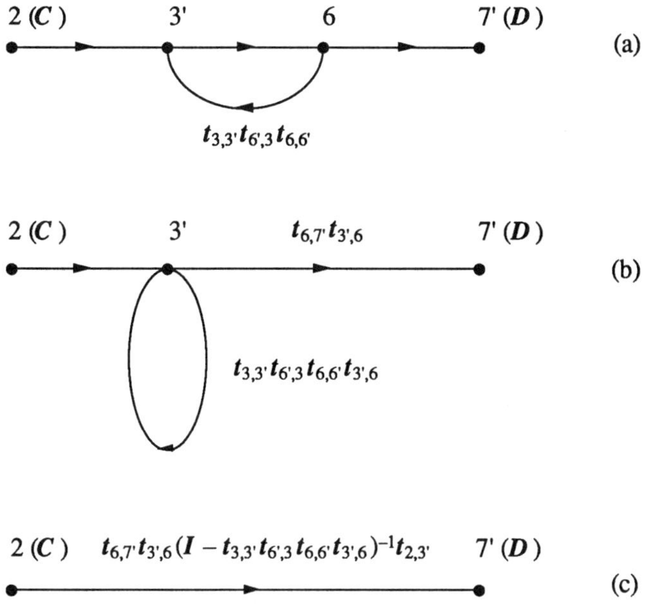

Figure 5.3 The reduction of the Fabry-Perot signal flow graph: (a) the signal flow graph after the application of the first reduction rule; (b) the signal flow graph after the elimination of node 6; (c) the signal flow graph after the elimination of the feedback branch and the application of the first reduction rule.

Furthermore, we assume that the waveguide connecting the two components is degenerate, and that its S matrix $\boldsymbol{F}$ is given by

$$\boldsymbol{F} = e^{2\pi i\nu\tau}\begin{pmatrix} \boldsymbol{0} & \boldsymbol{I} \\ \boldsymbol{I} & \boldsymbol{0} \end{pmatrix} \tag{5.3}$$

where τ is the delay time Ln/c corresponding to a waveguide of length L. In view of this, we have

$$\boldsymbol{t}_{2,3'} = \boldsymbol{t}_{6,7'} = \boldsymbol{t}$$

$$\boldsymbol{t}_{3',6} = \boldsymbol{t}_{6',3} = e^{2\pi i\nu\tau}\boldsymbol{I}$$

$$\boldsymbol{t}_{3,3'} = \boldsymbol{t}_{6,6'} = \boldsymbol{r}$$

From Figure 5.3(c) we find

$$t_{2,7'} = e^{2\pi i\nu\tau} t(I - e^{4\pi i\nu\tau} r^2)^{-1} t \tag{5.4}$$

If we assume that the matrices $\boldsymbol{r}$ and $\boldsymbol{t}$ are scalar (i.e., that $\boldsymbol{r} = r\boldsymbol{I}$ and $\boldsymbol{t} = t\boldsymbol{I}$), we obtain the familiar expression for the transfer matrix $\boldsymbol{J}$ of the Fabry-Perot interferometer [1]:

$$\boldsymbol{J} = t_{2,7'} = \frac{e^{2\pi i\nu\tau} t^2}{1 - e^{4\pi i\nu\tau} r^2} \boldsymbol{I} \tag{5.5}$$

The power transfer matrix is defined as $\lfloor \boldsymbol{J} \rfloor^2$ (Eq. (2.16)):

$$\lfloor \boldsymbol{J} \rfloor^2 = \frac{|t|^4}{1 - e^{4\pi i\nu\tau} r^2 - e^{-4\pi i\nu\tau} r^{*2} + |r|^4} \boldsymbol{I}$$

where r^* is the complex conjugate of r. Now let $r = |r|e^{i\theta}$:

$$\lfloor \boldsymbol{J} \rfloor^2 = \frac{(1 - |r|^2)^2}{(1 - |r|^2)^2 + 2|r|^2[1 - \cos(4\pi\nu\tau + 2\theta)]} \boldsymbol{I} \tag{5.6}$$

where we have assumed that the mirrors conserve power, so that

$$|t|^2 + |r|^2 = 1$$

Equation (5.6) can also be written in the form

$$\lfloor \boldsymbol{J} \rfloor^2 = \left[1 + \frac{4|r|^2}{(1 - |r|^2)^2} \sin^2(2\pi\nu\tau + \theta)\right]^{-1} \boldsymbol{I} \tag{5.7}$$

The numerical coefficient on the right-hand side is the familiar Fabry-Perot power transmission for polarized light [1]. Note that subwavelength changes in the waveguide length affect the values of both $\boldsymbol{J}$ and $\lfloor \boldsymbol{J} \rfloor^2$. Since the optical wavelength is of the order of magnitude of a micron, geometrical distortions on the submicron scale affect the transmission of the Fabry-Perot system.

5.3 THE RECIRCULATING LOOP

Another common optical network structure is the recirculating loop, shown in Figure 5.4(a). Its applications include, among others, signal processing [2] and fiber resonator construction [3].

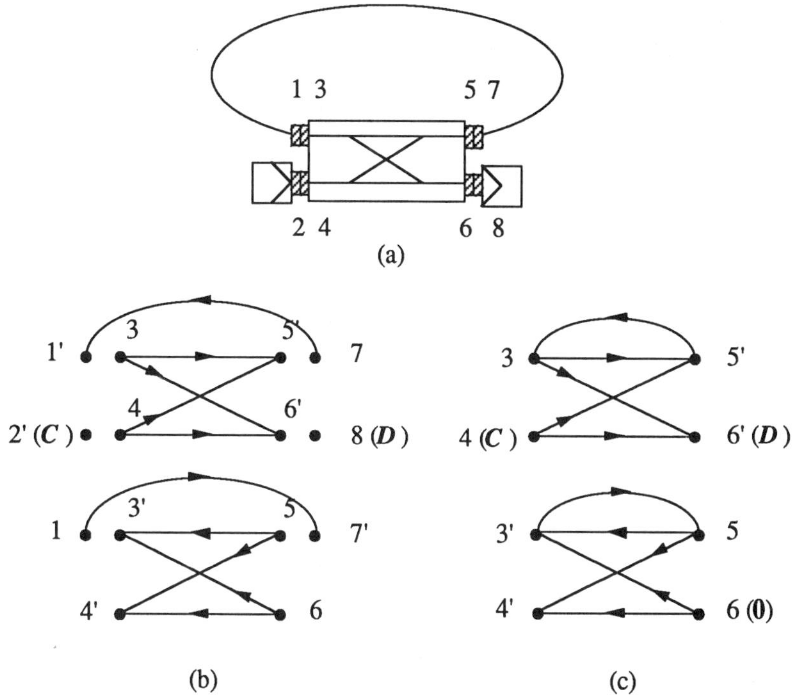

Figure 5.4 The physical scheme of the recirculating loop and its signal flow graph: (a) the physical scheme; (b) the signal flow graphs of the involved components before the connection; (c) the signal flow graph after the connection. The node variables associated with the lower part vanish, so it can be ignored.

In Figure 5.4(b) we show the unconnected signal flow graph of the recirculating loop, using again the simplified representation of the detector and the source. To represent the coupler we have used its disjoint signal flow graph form, as shown in Figure 4.10. Applying the signal flow graph connection rule, we obtain the graph shown in Figure 5.4(c). This graph consists of two separated parts. The lower part has only one source, namely node 6, and the value of its variable is **0**. Therefore, all node variables of this portion vanish, and it can be ignored. We could also have reached this conclusion by simple reasoning: since it is assumed that there are no reflections at the waveguide and the detector, the optical field propagates in the coupler from left to right only. Therefore, all the node variables of the the lower part of the coupler signal flow graph, which represents fields propagating in the opposite direction, have to vanish.

The solution of the upper part of the graph proceeds as follows. Applying the star elimination rule to node 5', we obtain the graph in Figure 5.5(a), and eliminating the resulting feedback branch we arrive at Figure 5.5(b). Applying the first reduction rule we obtain Figure 5.5(c), and the final reduction shown in Figure 5.5(d) is achieved by applying the parallel branch combination rule.

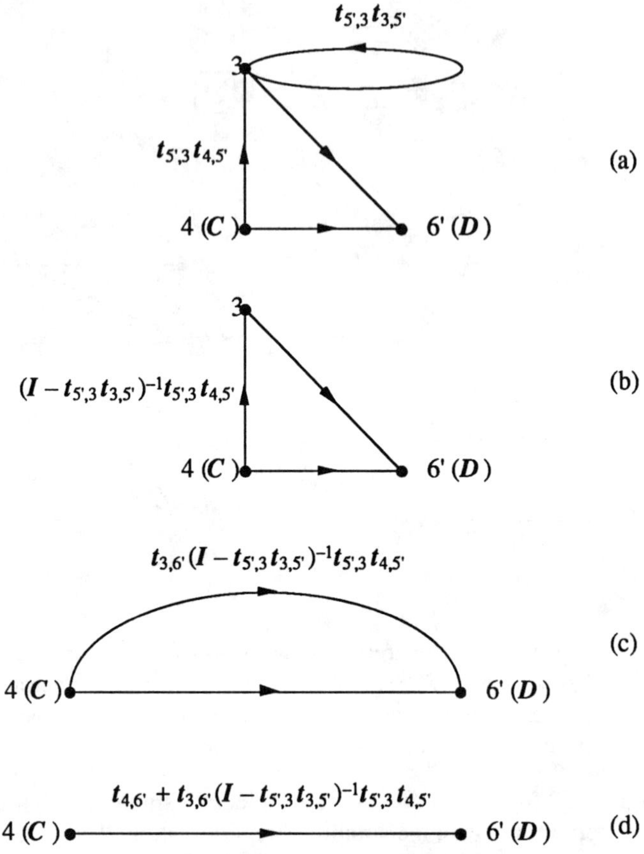

Figure 5.5 The reduction of the recirculating loop signal flow graph: (a) the signal flow graph after the elimination of the star node 5' (fourth reduction rule); (b) the signal flow graph after the elimination of the feedback branch (third reduction rule); (c) the signal flow graph after the combination of the serial branches (first reduction rule); (d) the final reduction following the combination of the parallel branches (second reduction rule).

To derive a more explicit expression for the transfer matrix J of the recirculating loop, we will assume for simplicity that the coupler is polarization-preserving. By also assuming that the waveguide is polarization-preserving, we may restrict the treatment to a single polarization mode and deal with scalars rather then Jones matrices, as explained in Chapter 3. In addition, we assume that the coupler is directional, reciprocal, and preserves power. Consequently, its S-submatrix s_C for a given polarization mode is of the general form shown in (3.14). Explicitly, using the representation (3.11), we have

$$s_C = \begin{pmatrix} 0 & 0 & e^{i\phi}\cos\theta & e^{i(\pi-\psi)}\sin\theta \\ 0 & 0 & e^{i\psi}\sin\theta & e^{-i\phi}\cos\theta \\ e^{i\phi}\cos\theta & e^{i\psi}\sin\theta & 0 & 0 \\ e^{i(\pi-\psi)}\sin\theta & e^{-i\phi}\cos\theta & 0 & 0 \end{pmatrix} \quad (5.8)$$

In this S-submatrix, the numbers of the rows and the columns correspond to the port numbers shown in Figure 3.3. Note also that we have suppressed the common phase $e^{i\alpha}$. In all our examples, its only effect is to add a constant phase to the phase of the the transfer function.

Referring to (5.6) and to Figure 3.3, we obtain

$$t_{4,6'} = e^{-i\phi}\cos\theta, \qquad t_{3,6'} = e^{i(\pi-\psi)}\sin\theta$$

$$t_{5',3} = e^{2\pi i\nu\tau}, \qquad t_{3,5'} = e^{i\phi}\cos\theta$$

$$t_{4,5'} = e^{i\psi}\sin\theta$$

where τ is as usual the delay time introduced by the waveguide connecting ports 3 and 5. Using these values for the transmissions in Figure 5.5(c), we obtain the following expression for the transfer function J:

$$J = t_{4,6'} = e^{-i\phi}\cos\theta - \frac{e^{2\pi i\nu\tau}\sin^2\theta}{1 - e^{2\pi i\nu\tau + i\phi}\cos\theta}$$

With some algebra, this expression can be simplified to

$$J = -\,e^{2\pi i\nu\tau}\,\frac{1 - e^{-i(2\pi\nu\tau+\phi)}\cos\theta}{1 - e^{i(2\pi\nu\tau+\phi)}\cos\theta} \quad (5.9)$$

It is interesting to note that the power transfer function $|J|^2$ is, in this case, constant, and has the value of 1. This fact is of course a direct result of our assumption that both the coupler and the waveguide are power-preserving.

5.4 THE GUIDED-WAVE MACH-ZEHNDER INTERFEROMETER

The physical scheme of the guided-wave version of the Mach-Zehnder interferometer is shown in Figure 5.6(a).

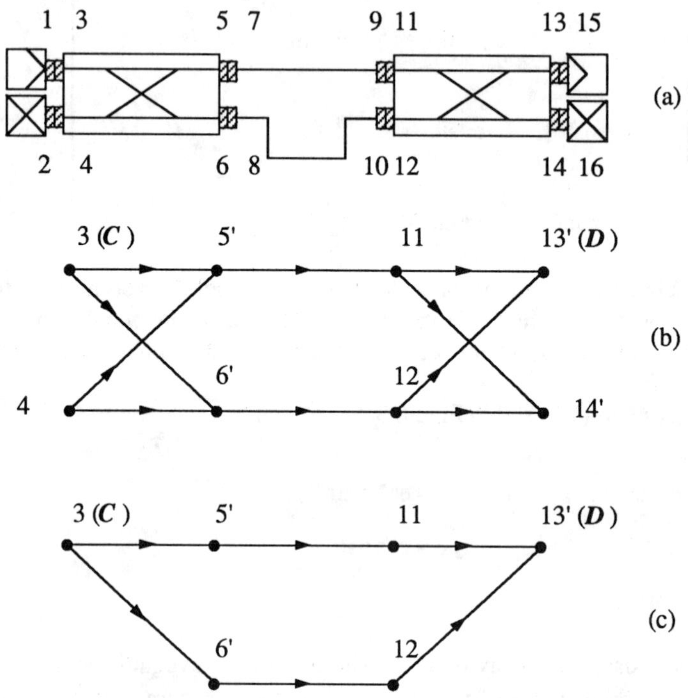

Figure 5.6 The derivation of the signal flow graph corresponding to a guided-wave Mach-Zehnder interferometer: (a) the physical scheme; (b) the signal flow graph after the connection; (c) the signal flow graph after the removal of the irrelevant nodes and branches.

Assuming that there are no reflections at the source, the detector, and the terminations, it can be inferred that the optical fields propagate throughout the network from left to right only. Therefore, referring to Figure 4.10, it is permissible to use only the upper part of the coupler signal flow graph. At this time we skip the unconnected stage, and directly show the connected signal flow graph in Figure 5.6(b). In this signal flow graph, we can ignore nodes 4 and 14'; the former is a null source, and the latter is a sink that represents the field incident on the right termination. After the removal of all nonessential nodes and branches from the graph shown in Figure 5.6(b), we get the final signal flow graph shown in Figure 5.6(c).

The reduction of the Mach-Zehnder signal flow graph is straightforward. Successively applying the first reduction rule to the appropriate segments, we arrive at the signal flow graph shown in Figure 5.7(a). Applying now the second reduction rule to this graph, we get the final form shown in Figure 5.7(b).

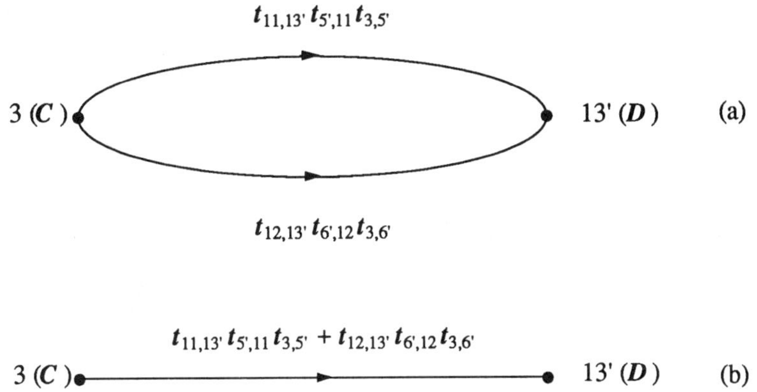

Figure 5.7 The reduction of the Mach-Zehnder signal flow graph: (a) the signal flow graph after the application of the first reduction rule; (b) the result of a successive application of the second reduction rule.

Let us assume again for simplicity that all the components that comprise the network are polarization-preserving. With reference to (5.6) and Figure 3.3, we infer the following values for the various transmissions appearing in Figure 5.6:

$$t_{11,13'} = \exp(i\phi_2)\cos\theta_2, \qquad t_{12,13'} = \exp(i\psi_2)\sin\theta_2$$

$$t_{5',11} = \exp(2\pi i\nu\tau_1), \qquad t_{6',12} = \exp(2\pi i\nu\tau_2)$$

$$t_{3,5'} = \exp(i\phi_1)\cos\theta_1, \qquad t_{3,6'} = \exp[i(\pi - \psi_1)]\sin\theta_1$$

In these relations, the subscripts 1 and 2 on the coupler parameters correspond to the left and the right coupler, respectively. The time delays τ_1 and τ_2 correspond to the upper and the lower waveguides, respectively. Using these values of the transmissions in the expression for the transfer function J as derived in Figure 5.7(b), we obtain

$$J = \exp[i(2\pi\nu\tau_1 + \phi_1 + \phi_2)]\cos\theta_1\cos\theta_2 - \exp[i(2\pi\nu\tau_2 + \psi_2 - \psi_1)]\sin\theta_1\sin\theta_2 \quad (5.10)$$

5.5 THE GUIDED-WAVE MICHELSON INTERFEROMETER

The physical scheme of the guided-wave version of the Michelson interferometer is shown in Figure 5.8(a). In this version, the conventional beam splitter is replaced by a directional coupler. The signal flow graph of this setup after the connection is shown in Figure 5.8(b). Before drawing this signal flow graph we have removed the nonessential nodes 3' and 4; the former is a null source, and the latter is a sink that represents the field incident on the source. Note that since, in this case, the power in the coupler flows in both directions, it is more convenient to use for its representation the signal flow graph shown in Figure 4.9. The properties of the signal flow graph are determined by the interconnections between the various nodes, not by their geometrical layout. In fact, the various nodes can be arbitrarily displaced as long as the connections between them remain intact. This operation can be used to simplify the geometrical layout of signal flow graphs in cases in which they seem to be unnecessarily complicated. The simplified version of Figure 5.8(b) is shown in Figure 5.9. Although these two graphs look geometrically very different, algebraically they are completely equivalent, since their node interconnection structure is identical.

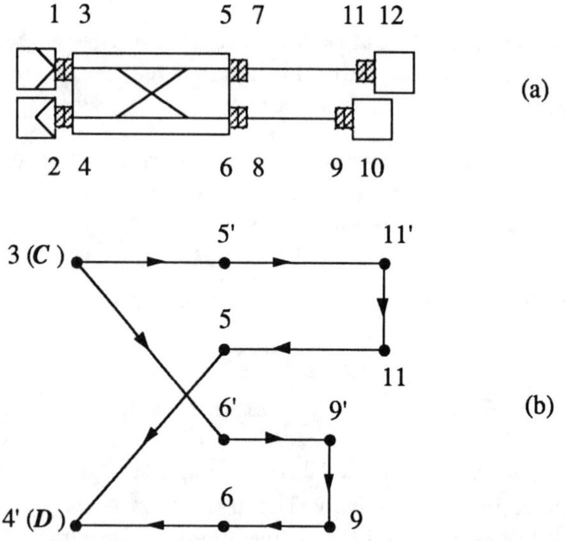

Figure 5.8 The derivation of the signal flow graph corresponding to a guided-wave Michelson interferometer: (a) the physical scheme; (b) the signal flow graph after the connection and the removal of the irrelevant nodes and branches.

The resulting signal flow graph is very similar to the signal flow graph of the guided-wave Mach-Zehnder interferometer, as shown in Figure 5.6(c), and the reduction process of both follows essentially the same lines. Applying the first and second reduction rules, we find the following expression for the transfer matrix $\boldsymbol{J}$ of the guided-wave Michelson interferometer:

$$\boldsymbol{J} = t_{3,4'} = t_{5,4'}t_{11,5}t_{11',11}t_{5',11'}t_{3,5'} + t_{6,4'}t_{9,6}t_{9',9}t_{6',9'}t_{3,6'}$$

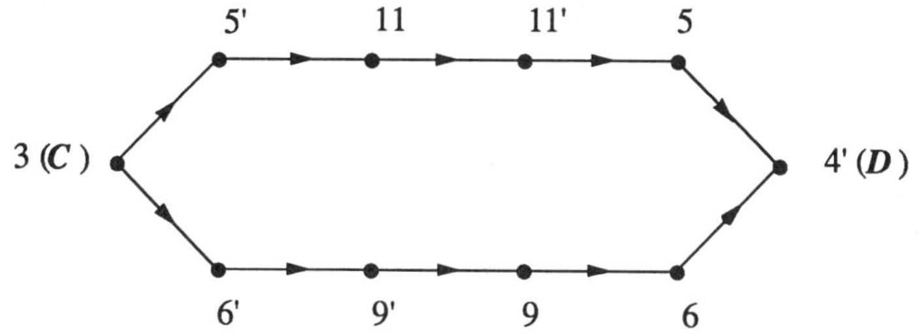

Figure 5.9 The simplified signal flow graph layout of the guided-wave Michelson interferometer.

As before, assuming a perfectly directional, polarization- and power-preserving and reciprocal coupler, the following values for the scalar transmissions are involved:

$$\begin{aligned}
t_{5,4'} &= e^{i\psi}\sin\theta, & t_{6,4'} &= e^{-i\phi}\cos\theta \\
t_{11,5} &= e^{2\pi i\nu\sigma}, & t_{9,6} &= e^{2\pi i\nu\tau} \\
t_{11',11} &= r_1, & t_{9',9} &= r_2 \\
t_{5',11'} &= e^{2\pi i\nu\sigma}, & t_{6',9'} &= e^{2\pi i\nu\tau} \\
t_{3,5'} &= e^{i\phi}\cos\theta, & t_{3,6'} &= e^{i(\pi-\psi)}\sin\theta
\end{aligned} \quad (5.11)$$

In (5.9) we have used the coupler S-submatrix parametrization of (5.6). Referring to Figure 5.8, the time delays σ and τ correspond to the upper and the lower interferometer arms, respectively. The reflection coefficients r_1 and r_2 correspond to the upper and the lower mirrors, respectively. Using these values for the transmissions, we obtain the following expression for the transfer function J of the guided-wave Michelson interferometer:

$$J = \frac{1}{2}\sin 2\theta[e^{i(4\pi\nu\sigma+\psi+\phi)}r_1 - e^{i(4\pi\nu\tau-\psi-\phi)}r_2] \quad (5.12)$$

It is worthwhile to note that the transfer function depends only on the sum of the angles ψ and ϕ, not on each of them individually.

5.6 THE ALGEBRAIC SOLUTION OF THE TIME-INDEPENDENT NETWORK PROBLEM

The solution of realistic networks by purely graphical means is frequently impractical because of the complexity of the operations that may be required to accomplish such a task. However, once the signal flow graph of the network is derived, it is an easy matter to write down the network equations and formulate the problem in a linear-algebra canonical form, which can be subsequently submitted to the now readily available machines for a numerical solution.

Let us denote the star variables with $\boldsymbol{x}_1, \boldsymbol{x}_2, ..., \boldsymbol{x}_L$, the source variables with $\boldsymbol{c}_1, \boldsymbol{c}_2, ..., \boldsymbol{c}_M$ and the sink variables with $\boldsymbol{d}_1, \boldsymbol{d}_2, ..., \boldsymbol{d}_N$. We remind the reader that each variable is a two-component vector (a Jones vector). Schematically, the signal flow graph has the form shown in Figure 5.10.

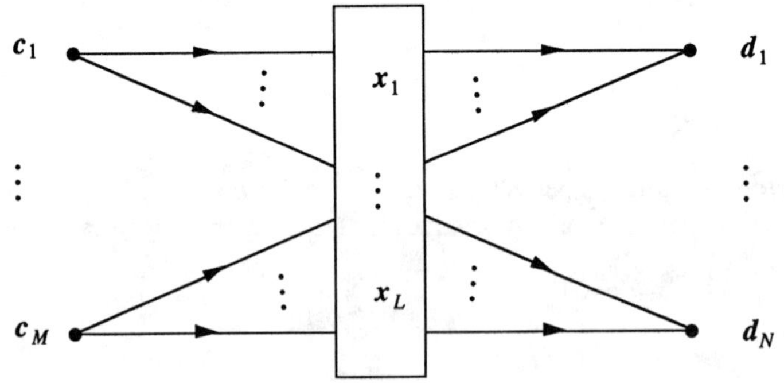

Figure 5.10 A schematic signal flow graph of a general network.

Using the signal flow graph association rule, we can derive a $2L\times2L$ matrix $\boldsymbol{R}$ and an $2M\times2L$ matrix $\boldsymbol{P}$ so that

$$\tilde{\boldsymbol{x}} = \boldsymbol{R}\tilde{\boldsymbol{x}} + \boldsymbol{P}\tilde{\boldsymbol{c}} \tag{5.13}$$

where

$$\tilde{\boldsymbol{x}} = \begin{pmatrix} \boldsymbol{x}_1 \\ \cdot \\ \cdot \\ \cdot \\ \boldsymbol{x}_L \end{pmatrix}, \qquad \tilde{\boldsymbol{c}} = \begin{pmatrix} \boldsymbol{c}_1 \\ \cdot \\ \cdot \\ \cdot \\ \boldsymbol{c}_M \end{pmatrix}$$

The formal solution of (5.13) is

$$\tilde{x} = (I - R)^{-1} P \tilde{c} \tag{5.14}$$

An additional set of equations that can be derived from Figure 5.10 is

$$\tilde{d} = Q \tilde{x} \tag{5.15}$$

where Q is a $2N \times 2L$ matrix and

$$\tilde{d} = \begin{pmatrix} d_1 \\ \cdot \\ \cdot \\ \cdot \\ d_N \end{pmatrix}$$

Equations (5.14) and (5.15) allow us to express the values of the sink variables $\tilde{d}$ in terms of the source variables $\tilde{c}$:

$$\tilde{d} = Q(I - R)^{-1} P \tilde{c} \tag{5.16}$$

Equation (5.16) may be regarded as a formal solution of the network problem. We see that a solution can be derived only if $I - R$ is nonsingular. The matrix $(I - R)^{-1}$ may be computed numerically.

It is worthwhile to note that in the time domain, (5.13) becomes, in general, a set of *integral* equations for the node variables (vector functions) $\tilde{x}$. The most straightforward way to solve these equations is by using the Fourier transform method, which is essentially a transformation to the frequency domain. This is another motivation to use the frequency-domain representation in the first place for the network analysis.

5.7 SUMMARY

This chapter explains the most elementary application of network analysis formalism: the analysis of time-independent networks. We started with the definition of the network algebra for the time-independent case. In the frequency domain, the product of two transmissions is reduced simply to the familiar algebraic product of two matrices. On the other hand, in the time domain this operation involves the computation of the convolution of two matrix functions. The summation operation of two transmissions is identical to the algebraic summation of two matrices in both the frequency and the time-domain representations.

The method has been illustrated by considering four explicit examples: the Fabry-Perot, the recirculating loop, the Mach-Zehnder, and the Michelson interferometer systems. Although we have in mind here guided-wave systems, the analysis is also valid for the fundamental mode of the corresponding bulk-optics interferometric systems.

In all four cases we have started the analysis with the derivation of the signal flow graphs. We proceeded with the reduction of the graphs by applying the four graphical reduction rules derived in the previous chapter. This led us to the derivation of the corresponding transfer matrices. The graphical reduction method is convenient for simple systems. In more complex systems it is necessary to resort to numerical methods, although simple reductions can sometimes be applied even in complex systems in order to simplify the numerical treatment. The network equations in the frequency domain are simply a set of linear algebraic equations, which (if nonsingular) can be solved by well-known numerical algorithms.

References

[1] Born, M. and E. Wolf, *Principles of Optics*, Oxford: Pergamon Press, Sixth Edition, 1980, p. 325.

[2] Moslehi B., J.W. Goodman, M. Tur and H.J. Shaw, "Fiber-Optic Lattice Signal Processing," *Proc. IEEE*, Vol. 72, No. 7, 1984, pp. 909-930.

[3] Zhang F. and J.W.Y. Lit, "Direct-Coupling Single-Mode Fiber Ring Resonator," *J. Opt. Soc. Am. A*, Vol. 5, No. 8, 1988, pp. 1347-1355.

Chapter 6
The Analysis of Networks That Are Periodic in Time

Time-dependent components are very common in modern optical networks. Communication networks use modulators to impress information on the optical carrier, sensors contain elements that change their optical properties as a result of an external influence, and heterodyne receivers use frequency shifters to generate the RF beat frequency at the detector. All these important applications cannot be treated properly by the purely time-independent formalism developed in Chapters 4 and 5. For the analysis of time-dependent networks, an appropriate extension of the time-independent formalism is required.

It seems that no generally useful tools can be developed for the treatment of networks containing components with an arbitrary time dependence. In such cases there is no advantage in using the frequency domain representation, and the solution of the network problem requires the inversion of a set of integral equations, both in the frequency and the time domains. However, certain general and useful tools can be developed for the special (and important) case of networks containing time-periodic components (components with a periodic time dependence). These tools are most appealing when there is either only one time-periodic component in the network, or all the components have the same fundamental period. In principle, networks containing components with commensurate fundamental periods can be treated with these tools as well by finding a common (longer than the fundamental) period, as outlined at the end of this chapter. This manipulation is, however, somewhat artificial, since the common period may be unrealistically long, particularly if a large number of components is involved.

The treatment of the time-periodic case is similar to the treatment of the time-independent case. The solution process is started with the derivation of the signal flow graph of the network, using exactly the same rules that we have used in the time-independent case. As before, we can proceed from this stage either by deriving the network equations and solving them numerically, or by attempting to reduce the graph using the signal flow graph reduction rules. However, the network algebra in the frequency domain is now different. What amounted to simple matrix-vector or matrix-matrix products in the time-independent case becomes now an infinite sum of such products. In the time domain, the network equations are a set of integral equations, as before; in the frequency domain, the network equations become, in general, a set of "sum" equations (the discrete analog to integral equations) instead of the simple algebraic equations that we had in the time-independent case.

In spite of these facts, the technique for the combination of two branches connected in series and in parallel (the first and the second reduction rules) is rather simple. The difficulty starts when we encounter a feedback branch. The equation associated with a node connected to a feedback branch is a sum equation in the frequency domain, and an integral equation in the time domain. In either case, it is not possible to offer a general solution to this equation, as we did in the frequency domain for the time-independent case. We do suggest, however, an approximate approach using the Neumann series, which may be applied in many cases of interest.

6.1 NETWORK ALGEBRA RULES FOR CYCLIC TRANSMISSIONS WITH IDENTICAL PERIODS

6.1.1 Preservation of the Periodic Time Dependence Upon Addition and Multiplication

The formulation of a network algebra for transmissions with identical periods is possible because the result of a sum or a product of such transmissions is also a periodic transmissions with the same period. The proof of this assertion is rather straightforward. Let k be an arbitrary integer. From (4.4) it follows that

$$\mathbf{J}(t_2 + kT, t_1 + kT) = \mathbf{J}_1(t_2 + kT, t_1 + kT) + \mathbf{J}_2(t_2 + kT, t_1 + kT)$$

If $\mathbf{J}_1$ and $\mathbf{J}_2$ are both cyclic with a period T, then, according to (2.18),

$$\mathbf{J}(t_2, t_1) = \mathbf{J}(t_2 + kT, t_1 + kT) \tag{6.1}$$

which proves that $\mathbf{J}$ is also cyclic with a period T.

Let us consider now the product of two transmissions. From (4.3) it follows that

$$\mathbf{J}(t_3 + kT, t_1 + kT) = \int dt_2\, \mathbf{J}_2(t_3 + kT, t_2)\mathbf{J}_1(t_2, t_1 + kT)$$

where k is again an arbitrary integer. Introducing a new integration variable $t = t_2 - kT$, we obtain

$$\begin{aligned} \mathbf{J}(t_3 + kT, t_1 + kT) &= \int dt \mathbf{J}_2(t_3 + kT, t + kT)\mathbf{J}_1(t + kT, t_1 + kT) \\ &= \int dt \mathbf{J}_2(t_3, t)\mathbf{J}_1(t, t_1) \\ &= \mathbf{J}(t_3, t_1) \end{aligned} \tag{6.2}$$

which proves that **J** is cyclic with a period T. Note that the above proofs are also valid if $\mathbf{J}_1$ or $\mathbf{J}_2$ (or even both) are time-invariant (i.e., represent time-independent components). Another way to recognize this fact is to recall that time-invariant Jones matrices may be regarded as cyclic with an arbitrary period.

6.1.2 The Network Algebra Addition Operation in the Frequency Domain

The network algebra sum and product operations in the time domain for the time-dependent and time-independent cases are basically identical. However, the corresponding operations in the frequency domain are significantly modified. In Section (2.4) we saw that the frequency-domain representation of a cyclic Jones matrix $\boldsymbol{J}$ is a (perhaps infinite) set of matrices $\boldsymbol{j}_n$, which were called the Fourier expansion coefficients of $\boldsymbol{J}$. Thus, in the frequency domain, the branch transmissions should be regarded as *sets* of matrices rather than as single matrices as we had in the time-independent case. To establish the network algebra in the frequency domain, we have to define the relationship between the two matrix sets involved in the summation and the product operations and the resultant matrix set.

The derivation of the frequency-domain sum of cyclic Jones matrices is straightforward. Let $\mathbf{U}(t_2,t_1)$ and $\mathbf{V}(t_2,t_1)$ be two such matrices, with the expansions (2.21):

$$\mathbf{U}(t_2,t_1) = \sum_n \exp(-2\pi i n f_0 t_2) \int d\nu \boldsymbol{u}_n(\nu) \exp[-2\pi i \nu (t_2 - t_1)] \tag{6.3}$$

$$\mathbf{V}(t_2,t_1) = \sum_n \exp(-2\pi i n f_0 t_2) \int d\nu \boldsymbol{v}_n(\nu) \exp[-2\pi i \nu (t_2 - t_1)] \tag{6.4}$$

where $f_0 = 1/T$. The sum

$$\mathbf{W}(t_2,t_1) = \mathbf{U}(t_2,t_1) + \mathbf{V}(t_2,t_1)$$

has the expansion

$$\mathbf{W}(t_2,t_1) = \sum_n \exp(-2\pi i n f t_2)\int d\nu [\boldsymbol{u}_n(\nu) + \boldsymbol{v}_n(\nu)]\exp[-2\pi i \nu(t_2 - t_1)]$$

so that the Fourier expansion coefficients $\boldsymbol{w}_n(\nu)$ of $\mathbf{W}(t_2,t_1)$ are simply given by

$$\boldsymbol{w}_n(\nu) = \boldsymbol{u}_n(\nu) + \boldsymbol{v}_n(\nu) \tag{6.5}$$

Let us denote the set of the Fourier expansion coefficients of a cyclic Jones matrix **J** with $\tilde{j}$. In view of (6.5), we define the sum of two sets of Fourier expansion coefficients $\tilde{\boldsymbol{u}}$ and $\tilde{\boldsymbol{v}}$ as a set whose kth element is the sum of the kth elements of $\tilde{\boldsymbol{u}}$ and $\tilde{\boldsymbol{v}}$. Using this definition, we can write (6.5) as an equality between *sets* of matrices:

$$\tilde{\boldsymbol{w}}(n) = \tilde{\boldsymbol{u}}(n) + \tilde{\boldsymbol{v}}(\nu) \tag{6.6}$$

6.1.3 The Network Algebra Product Operation in the Frequency Domain

We turn now to the computation of the Fourier expansion coefficients corresponding to the product of two cyclic Jones matrices. Let

$$\mathbf{W}(t_3,t_1) = \int dt_2 \mathbf{V}(t_3,t_2)\mathbf{U}(t_2,t_1)$$

Using the expansions (6.3) and (6.4), we find

$$\mathbf{W}(t_3,t_1) = \sum_{n,m}\int d\mu d\nu \boldsymbol{v}_n(\mu)\boldsymbol{u}_m(\nu)\exp\{2\pi i[\nu t_1 - t_3(\mu + n f_0)]\}$$
$$\cdot\int dt_2 \exp[2\pi i t_2(-m f_0 + \mu - \nu)]$$

This expression can also be cast in the form

$$\mathbf{W}(t_3,t_1) = \sum_{m,k} \exp(-2\pi i k f_0 t_3)\int d\nu \boldsymbol{v}_{k-m}(\nu + m f_0)\boldsymbol{u}_m(\nu)\exp[-2\pi i \nu(t_3 - t_1)] \tag{6.7}$$

The Fourier representation of $\mathbf{W}(t_3,t_1)$ is

$$\mathbf{W}(t_3,t_1) = \sum_n \exp(-2\pi i n f_0 t_3)\int d\nu \boldsymbol{w}_n(\nu)\exp[-2\pi i \nu(t_3 - t_1)]$$

Comparing this expansion and (6.7), we obtain

$$[\tilde{v}\ \tilde{u}]_k(\nu) = \sum_m v_{k-m}(\nu + mf_0)u_m(\nu) \tag{6.8}$$

Note that (6.8) is not an equality between sets as is (6.6), but rather between set elements. We see that for each Fourier expansion coefficient of a product we generally have to compute an infinite sum. Still, a sum is usually easier to evaluate than an integration, which is necessary for the computation of the product in the time domain, as indicated in (6.1). It must be admitted, however, that, in general, the advantage of the frequency-domain representation is partially lost when cyclic Jones matrices are introduced. In fact, for the analysis of networks containing time-periodic components with an instantaneous time response, it is easier to use the time-domain representation and derive the Fourier expansion coefficients of the final result, if desired, by using (2.23).

6.1.4 Operations Between Time-Invariant and Cyclic Transmissions

Even if a given network contains time-dependent components, it is quite likely that many, if not most of the transmissions in its signal flow graph will be time-invariant. The combination of a time-invariant and a cyclic Jones matrix is a special case of the combination of two cyclic matrices considered above: with each time-invariant Jones matrix we associate a set of Fourier expansion coefficients consisting of only one nonvanishing element corresponding to $k = 0$ (2.22). However, in view of the common occurrence of operations in which one matrix is cyclic and one time-invariant, it is convenient to derive explicit expressions for this special case. If $\boldsymbol{J}(\nu)$ is the frequency-domain representation of a time-invariant Jones matrix, and $\tilde{\boldsymbol{u}}$ is a set of Fourier expansion coefficients of a cyclic Jones matrix, then we have for the sum

$$[\tilde{\boldsymbol{u}} + \tilde{\boldsymbol{J}}]_k(\nu) = \boldsymbol{u}_k + \boldsymbol{J}(\nu)\delta_{k,0} \tag{6.9}$$

and for the two possible products

$$[\tilde{\boldsymbol{u}}\ \tilde{\boldsymbol{J}}]_k(\nu) = \boldsymbol{u}_k(\nu)\boldsymbol{J}(\nu) \tag{6.10}$$

$$[\tilde{\boldsymbol{J}}\ \tilde{\boldsymbol{u}}]_k(\nu) = \boldsymbol{J}(\nu + kf_0)\boldsymbol{u}_k(\nu) \tag{6.11}$$

The signal flow graphs corresponding to (6.10) and (6.11) are shown in Figure 6.1. Note that in the computation of the kth Fourier expansion coefficient of the result, the argument of a time-invariant transmission following a cyclic transmission is raised by kf_0. In the special case in which $\boldsymbol{J}$ is independent of ν, we may use the simplified rule

$$[\tilde{J}\,\tilde{u}]_k(\nu) = J u_k(\nu) \tag{6.11a}$$

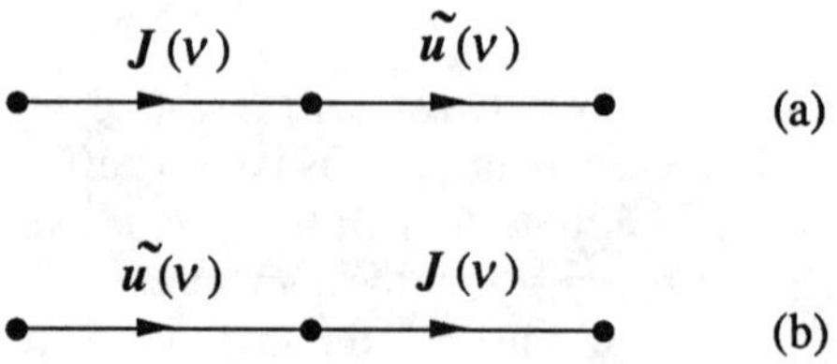

Figure 6.1 A time-invariant transmission J and a cyclic transmission having a set $\tilde{u}$ of Fourier expansion coefficients connected in series: (a) a cyclic transmission following a time-invariant transmission; (b) a time invariant transmission following a cyclic transmission.

6.2 SERIAL COMBINATION OF TRANSMISSIONS REPRESENTING AMPLITUDE MODULATORS, FREQUENCY SHIFTERS, AND PHASE MODULATORS (EXAMPLES)

6.2.1 The Transfer Matrix of an Instantaneous Modulator With Leads

In order to familiarize ourselves with the formalism presented in the previous section, let us consider a few simple examples. In these examples we will assume that the modulators consist of an intrinsically instantaneous device separated from the ports by two ideal and degenerate waveguide leads with delay times σ_1 and σ_2, as shown in Figure 6.2(a). To derive the transfer matrix of such a modulator, we insert in the original scheme two pairs of virtual ports (Figure 6.2(b)), and obtain the signal flow graph shown in Figure 6.2(c). We have assumed for simplicity that the modulating device does not give rise to reflections; thus, the signal flow graph is made up from two disconnected parts.

Let us consider the upper part of the signal flow graph. It consists of three transmissions $\mathbf{J}_1$, $\mathbf{J}_M$, and $\mathbf{J}_2$ connected in series. According to our assumptions, these transmissions are given by

$$\mathbf{J}_1(t_2,t_1) = \delta(t_2 - t_1 - \tau_1)$$

$$\mathbf{J}_M(t_2,t_1) = \mathbf{U}(t_2)\delta(t_2 - t_1)$$

$$\mathbf{J}_2(t_2,t_1) = \delta(t_2 - t_1 - \tau_2)$$

Combining these three transmissions by twice applying the rule for the combination of serial transmissions in the time domain (Eq. (6.1)), we find the following expression for the transmission $\mathbf{J}(t_2,t_1)$ of the upper part:

$$\mathbf{J}(t_2,t_1) = \mathbf{U}(t_2 - \tau_2)\delta(t_2 - t_1 - \tau)$$

where $\tau = \tau_1 + \tau_2$ is the total delay time. Since the matrix $\mathbf{U}(t)$ is periodic in time, a constant shift in its time argument results simply in an alteration of its cycle phase. Thus, without loss of generality, we are free to use for $\mathbf{J}(t_2,t_1)$ the simplified form

$$\mathbf{J}(t_2,t_1) = \mathbf{U}(t_2)\delta(t_2 - t_1 - \tau) \tag{6.12}$$

by adding an appropriate phase to the operation cycle of $\mathbf{U}(t)$. The Fourier expansion coefficients $\boldsymbol{j}_n$ of $\mathbf{J}$ are found to be

$$\boldsymbol{j}_n = e^{2\pi i \nu \tau}\boldsymbol{u}_n \tag{6.13}$$

where $\boldsymbol{u}_n$ is the discrete Fourier transform of the periodic matrix $\mathbf{U}$.

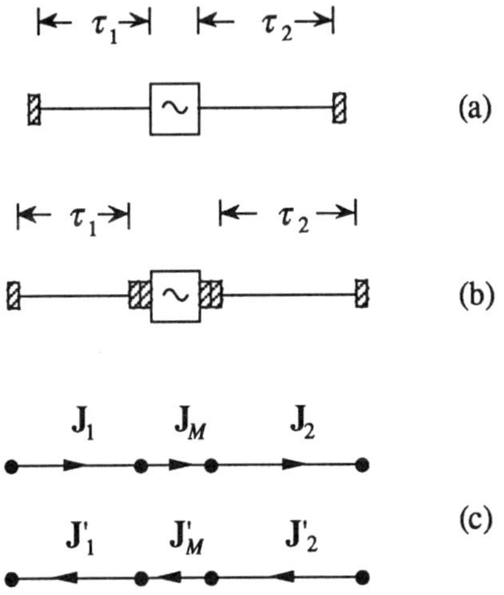

Figure 6.2 The modulator that will be used in the examples: (a) the physical scheme; (b) the physical scheme with the virtual ports; (c) the signal flow graph of the modulator.

Since reciprocity does not apply to time-dependent components, there is no general relation between the transmissions of the upper and the lower parts in Figure 6.2(c).

The physical scheme of two modulators connected in series is shown in Figure 6.3. Due to the assumption that these modulators do not give rise to reflections, the signal flow graph corresponding to this setup will also consist of two disjoint parts. In what follows, we will consider the upper part only.

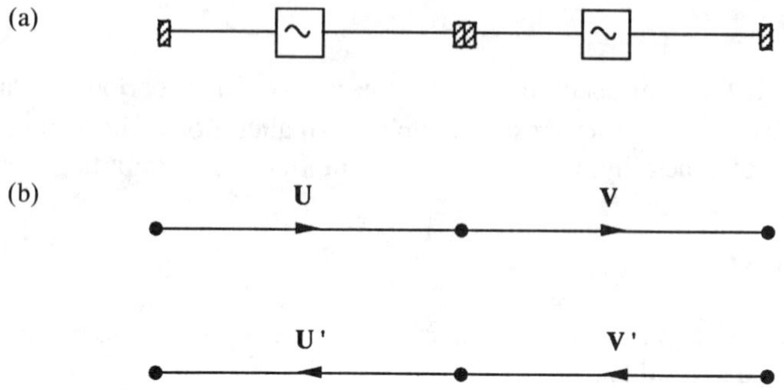

Figure 6.3 Two modulators connected in series: (a) the physical scheme; (b) The corresponding signal flow graph.

6.2.2 Two Amplitude Modulators in Series

We start with the case of two amplitude modulators of the type in (2.28) connected in series. Let the individual transfer matrices of these modulators be

$$\mathbf{U}(t_2,t_1) = \tfrac{1}{2}\,\boldsymbol{U}_0[1 + \cos(2\pi f_0 t_2 + \phi_1)]\,\delta(t_2 - t_1 - \tau_1)$$

$$\mathbf{V}(t_2,t_1) = \tfrac{1}{2}\,\boldsymbol{V}_0[1 + \cos(2\pi f_0 t_2 + \phi_2)]\,\delta(t_2 - t_1 - \tau_2)$$

where $\boldsymbol{U}_0$ and $\boldsymbol{V}_0$ are some constant matrices and τ_1 and τ_2 are the delay times associated with **U** and **V** respectively. Applying the multiplication rule in the time domain (Eq. (6.1)) we obtain for the product transmission **W**

$$\mathbf{W}(t_2,t_1) = \tfrac{1}{4}\,\boldsymbol{W}_0[1 + \cos(2\pi f_0 t_2 + \phi_2)][1 + \cos(2\pi f_0 t_2 + \phi_1')]\,\delta(t_2 - t_1 - \tau)$$

where

$$\boldsymbol{W}_0 = \boldsymbol{V}_0\boldsymbol{U}_0$$

$$\phi_1' = \phi_1 - 2\pi f_0 \tau_2$$

and τ is the total delay time $\tau_1 + \tau_2$. Using a well-known trigonometrical identity, we cast this relation in the form

$$\mathbf{W}(t_2,t_1) = \tfrac{1}{4} W_0[1 + \cos(2\pi f_0 t_2 + \phi_1') + \cos(2\pi f_0 t_2 + \phi_2) + \tfrac{1}{2}\cos(\phi_1' - \phi_2)$$
$$+ \tfrac{1}{2}\cos(4\pi f_0 t_2 + \phi_1' + \phi_2)]\delta(t_2 - t_1 - \tau)$$

The Fourier expansion coefficients of **W** are obtained by applying (6.13):

$$w_k(\nu) = \tfrac{1}{4} W_0\{[1 + \tfrac{1}{2}\cos(\phi_1 - \phi_2)]\delta_{k,0} + \tfrac{1}{2}[\exp(-ik\phi_1') + \exp(-ik\phi_2)]\delta_{|k|,1}$$
$$+ \tfrac{1}{4}\exp[-ik(\phi_1' + \phi_2)]\delta_{|k|,2}\}e^{2\pi i\nu\tau}$$

6.2.3 Combination of Frequency Shifters in Series

We consider now the case of two frequency shifters of the type in (2.30) connected in series. Let the transmissions of the frequency shifters be

$$\mathbf{U}(t_2,t_1) = U_0\exp(-2\pi i f_0 t_2)\delta(t_2 - t_1 - \tau_1)$$

$$\mathbf{V}(t_2,t_1) = V_0\exp(-2\pi i f_0 t_2)\delta(t_2 - t_1 - \tau_2)$$

where we have "absorbed" the constant phases in U_0 and V_0. Applying the multiplication rule in the time domain, we obtain the following expression for the resulting transmission of two frequency shifters connected in series:

$$\mathbf{W}(t_2,t_1) = W_0\exp(-4\pi i f_0 t_2)\delta(t_2 - t_1 - \tau) \tag{6.14}$$

Thus, the combination in series of two frequency shifters with frequency f_0 acts as a frequency shifter with frequency $2f_0$. The Fourier expansion coefficients of **W** are

$$w_n = W_0 e^{2\pi i\nu\tau}\delta_{n,2} \tag{6.15}$$

From (6.15) it is easy to infer the expression for the Fourier expansion coefficients $w^{(N)}{}_k$ of N identical frequency shifters connected in series:

$$w^{(N)}{}_k = V_0^N e^{2\pi i\nu\tau}\delta_{N,k} \tag{6.16}$$

where τ is the total time delay of the N modulators.

6.2.4 Combination of Phase Modulators in Series

After examining these rather simple cases, we turn to the more interesting case of the combination of two phase modulators. Let the transmissions of the two phase modulators of the type in (2.32) be

$$\mathbf{U}(t_2,t_1) = U_0\exp[-i\gamma_1\sin(2\pi f_0 t_2 + \phi_1)]\delta(t_2 - t_1 - \tau_1)$$

$$\mathbf{V}(t_2,t_1) = V_0\exp[-i\gamma_2\sin(2\pi f_0 t_2 + \phi_2)]\delta(t_2 - t_1 - \tau_2)$$

Applying the multiplication rule for this case, we find the resulting transmission $\mathbf{W}(t_2,t_1)$:

$$\mathbf{W}(t_2,t_1) = W_0\exp[-i\gamma_1\sin(2\pi f t_2 + \phi_1) - i\gamma_2\sin(2\pi f t_2 + \phi_2)]\delta(t_2 - t_1 - \tau) \quad (6.17)$$

To establish an interesting interpretation of (6.17), let us define the complex modulation indices Γ_1 and Γ_2:

$$\Gamma_k = \gamma_k\exp(i\phi_k), \quad k = 1,2$$

We introduce now

$$\Gamma = \Gamma_1 + \Gamma_2 \quad (6.18)$$

Multiplying both sides of (6.18) by $\exp(2\pi i f_0 t)$ and taking the imaginary part, we obtain

$$\gamma\sin(2\pi f_0 t + \phi) = \gamma_1\sin(2\pi f_0 t + \phi_1) + \gamma_2\sin(2\pi f_0 t + \phi_2) \quad (6.19)$$

where γ and ϕ are defined by $\Gamma = \gamma e^{i\phi}$. This is recognized as the familiar addition rule of two AC signals with the same frequency and different phases and amplitudes. From (6.17) and (6.19) we obtain

$$\mathbf{W}(t_2,t_1) = W_0\exp[-i\gamma\sin(2\pi f_0 t_2 + \phi)]\delta(t_2 - t_1 - \tau) \quad (6.20)$$

Thus, the combination of two phase modulators in series acts as a new phase modulator whose complex modulation index is the sum of the complex modulation indices of the original ones. More generally, the combination of N phase modulators with complex modulation indices $\Gamma_1, \Gamma_2, \ldots, \Gamma_N$ results in an effective phase modulator whose complex modulation index Γ is given by

$$\Gamma = \sum_{k=1}^{N} \Gamma_k \tag{6.21}$$

The Fourier expansion coefficients of **W** can be derived from (6.13) and (2.33):

$$\mathbf{w}_n = W_0 e^{2\pi i \nu\tau - in\phi} J_n(\gamma) \tag{6.22}$$

Of particular interest is the case in which the moduli of Γ_1, Γ_2, ..., Γ_N are equal, but their sequential phases are shifted from each other by a constant amount $\Delta\phi$:

$$\gamma_1 = \gamma_2 = \ldots = \gamma_N$$

$$\phi_k = \phi_1 + (k - 1)\Delta\phi, \quad k = 2, \ldots, N$$

In this case, the complex modulation indices form a geometrical series, the sum of which is

$$\Gamma = \Gamma_1 \frac{e^{iN\Delta\phi} - 1}{e^{i\Delta\phi} - 1}$$

This relation can also be written as

$$\gamma = \gamma_1 F_N(\Delta\phi), \tag{6.23}$$

$$\phi = \phi_1 + \tfrac{1}{2}\Delta\phi(N - 1) \tag{6.24}$$

where the function $F_n(\Delta\phi)$ is defined by

$$F_N(\Delta\phi) = \frac{\sin(\frac{1}{2}N\Delta\phi)}{\sin(\frac{1}{2}\Delta\phi)} \tag{6.25}$$

The function $F_N(\Delta\phi)$ occurs in the description of a superposition of waves that are successively shifted in phase by a constant amount, as, for instance, in the description of the far field radiation pattern created by a diffraction grating illuminated by a plane wave [1]. This function exhibits extrema for $\Delta\phi_n = 2\pi n$, with

$$F_N(\Delta\phi_n) = (-1)^{n(N-1)} N \tag{6.26}$$

The width of the extremes is roughly π/N.

The transmission $\mathbf{W}$ of N serial phase modulators successively shifted by $\Delta\phi$ is therefore given by

$$\mathbf{W}(t_2,t_1) = \mathbf{W}_0\exp\{-i\gamma_1 F_N(\Delta\phi)\sin[2\pi f_0 t_2 + \phi_1 + \tfrac{1}{2}\Delta\phi(N-1)]\}\delta(t_2 - t_1 - \tau)$$

and its Fourier expansion coefficients $\mathbf{w}_n$ by

$$\mathbf{w}_n = W_0\exp\{2\pi i\nu\tau - ik[\phi_1 + \tfrac{1}{2}\Delta\phi(N-1)]\}J_n[\gamma_1 F_N(\Delta\phi)] \tag{6.27}$$

where τ is the total delay time of the N modulators. If this phase modulator operates on a monochromatic light of frequency ν_0, then the power P_n in the output component at the frequency $\nu_0 + nf_0$ is proportional to p_n, where p_n is given by

$$p_n = J_n^2[\gamma_1 F_N(\Delta\phi)] \tag{6.28}$$

Representative plots of p_n for n = 0, 1, and 2 as a function of $\Delta\phi$ for γ_1 = 1 and N = 10 are given in Figure 6.4.

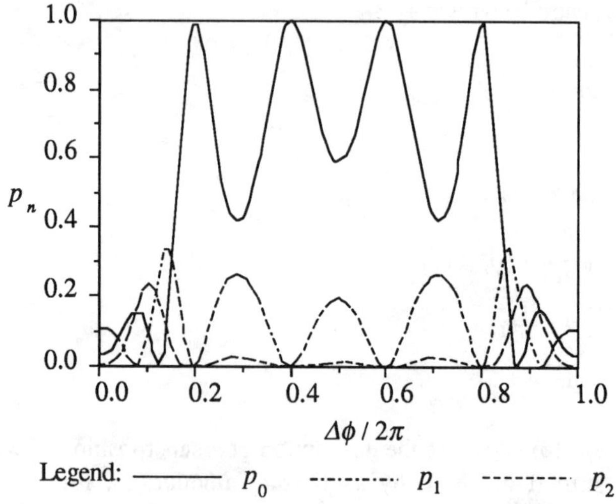

Figure 6.4 The normalized optical power p_n in the frequency component $\nu_0 + nf_0$ generated by 10 identical phase modulators successively shifted in phase by $\Delta\phi$.

6.3 THE MODULATED MACH-ZEHNDER INTERFEROMETER (EXAMPLE)

So far we have seen how to multiply or add individual branch transmissions. To illustrate how these operations can be used for the analysis of complete networks, we consider the simple example of the modulated Mach-Zehnder interferometer.

Suppose that we replace one of the waveguides in Figure 5.6(a) with a modulator of the type we have considered in the previous section, as shown in Figure 6.5.

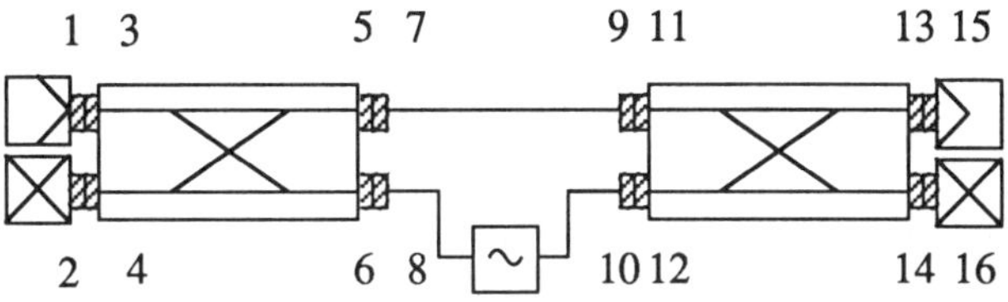

Figure 6.5 A guided-wave Mach-Zehnder interferometer with a modulated arm.

As we have mentioned, the introduction of a time-dependent component does not affect the derivation of the signal flow graph, nor its reduction. Consequently, we can use our earlier results shown in Figures 5.6 and 5.7 for the analysis of the present case. Referring to Figure 5.7(b), we therefore write the Fourier expansion coefficient set $\tilde{w}$ of the transfer matrix W of this network as

$$\tilde{w} = [\tilde{t}_{11,13'}\tilde{t}_{5',11}\tilde{t}_{3,5'}] + [\tilde{t}_{12,13'}\tilde{t}_{6',12}\tilde{t}_{3,6'}] \tag{6.29}$$

All transmissions in (6.29) are, in principle, sets of Fourier expansion coefficients. However, only $\tilde{t}_{6',12}$, the one that represents the modulator, is truly cyclic; all other transmissions are time-invariant. Denoting the kth Fourier expansion coefficient of $\tilde{t}_\alpha$ with $t_{\alpha;k}$ for all time-invariant transmissions $\tilde{t}_\alpha$, we have

$$t_{\alpha;k} = t_\alpha \delta_{k,0}$$

where t_α is simply the corresponding Jones matrix of the transmission when regarded as time-invariant (2.22). Using rules (6.9) to (6.11) for the combination of cyclic and time-invariant transmissions, we find that

$$w_k(\nu) = t_{11,13'}(\nu)t_{5',11}(\nu)t_{3,5'}(\nu)\delta_{k,0} + t_{12,13'}(\nu + kf_0)t_{6',12}(\nu)t_{3,6'}(\nu) \tag{6.30}$$

Equation (6.30) gives the general expression for the Fourier expansion coefficients of the transfer matrix of a guided-wave Mach-Zehnder interferometer. We note that the time-independent arm affects only the zeroth Fourier expansion coefficient w_0.

Let us assume now that (i) the upper interferometer arm is made up from an ideal, degenerate waveguide with a delay time τ_1, (ii) the Fourier expansion coefficients of the modulator are of the form (6.13):

$$t_{6',12;k} = \exp(2\pi i \nu \tau_2) u_k$$

where τ_2 is the delay time associated with it, and (iii) all other transmissions in (6.30) are independent of ν. Under these assumptions we have

$$w_k(\nu) = \exp(2\pi i \nu \tau_1) t_{11,13'} t_{3,5'} \delta_{k,0} + \exp(2\pi i \nu \tau_2) t_{12,13'} u_k t_{3,6'} \tag{6.31}$$

It can be seen that the two-exponential superposition frequency dependence familiar from the time-independent case is preserved in w_0 only (if u_0 does not vanish).

6.4 THE TREATMENT OF A FEEDBACK BRANCH

6.4.1 The General Procedure

The solution of the guided-wave Mach-Zehnder interferometer was rather simple since it did not contain feedback branches. In this section we consider the problem of the feedback branch elimination in the time-dependent case.

Let us consider again a signal flow graph containing a node attached to a feedback branch as shown in Figure 6.6. For simplicity we show only one incoming branch.

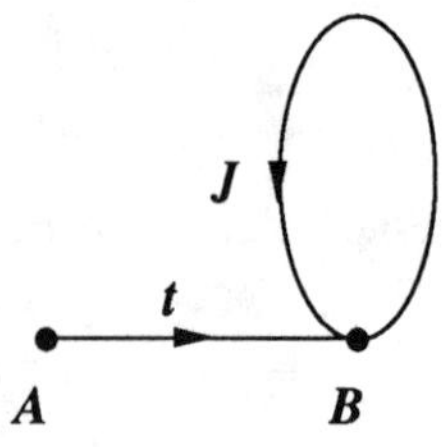

Figure 6.6 A signal flow graph containing a node attached to a feedback branch.

The equation associated with the node variable B is

$$B = JB + tA \tag{6.32}$$

In the time domain, (6.32) reads as

$$\mathbf{B}(t_2) = \int dt_1 \, [\mathbf{J}(t_2,t_1)\mathbf{B}(t_1) + \mathbf{t}(t_2,t_1)\mathbf{A}(t_1)]$$

Let us assume now that **J** is cyclic, and **t** is time-invariant. Under these assumptions, the frequency-domain representation of (6.32) reads

$$B(\nu) = \sum_n j_n(\nu - nf_0)B(\nu - nf_0) + t(\nu)A(\nu)$$

While there is no general solution to these equations, either in the frequency or in the time domains, there is a generally useful method to find successive approximations $B^{(n)}$ to the solution using the iterative process

$$B^{(n+1)} = JB^{(n)} + tA \tag{6.33}$$

starting with

$$B^{(0)} = tA \tag{6.34}$$

The series $B^{(n)}$ is called *Neumann series* [2]. Under certain conditions, the Neumann series converges to the desired solution. On physical grounds, one can expect the Neumann series to converge if the transmission J is lossy, i.e., if $I - [J^{\dagger}J]$ is a positive definite operator. On the other hand, if the feedback branch contains an active element that amplifies the signal, the Neumann series will generally not converge, and a different approach for the solution of (6.32) must be used, as in, for instance, the *Fredholm series* expansion [2].

If the Neumann series converges, then we can present a formal solution to the feedback branch problem in the form

$$B = \sum_{n=0}^{\infty} [J^n t]A \tag{6.35}$$

where J^n stands for the product of J with itself n times, and

$$J^0 = I \tag{6.36}$$

Formally, we may regard (6.35) as a sum of an infinite geometrical series, and replace it with the well-known formula to obtain

$$\boldsymbol{B} = [(\boldsymbol{I} - \boldsymbol{J})^{-1} t\boldsymbol{A}] \tag{6.37}$$

Alternatively, (6.35) may be regarded as a "Taylor expansion" of (6.37). Formally, (6.37) is identical to what we would have obtained by the application of the feedback branch elimination rule of the time-independent case. However, in spite of its appealing simplicity, (6.37) has little use in practice, where we normally use a truncated version of (6.35) to obtain an approximate solution.

It is instructive to present the signal flow graph corresponding to (6.35). This signal flow graph is shown in Figure 6.7 below.

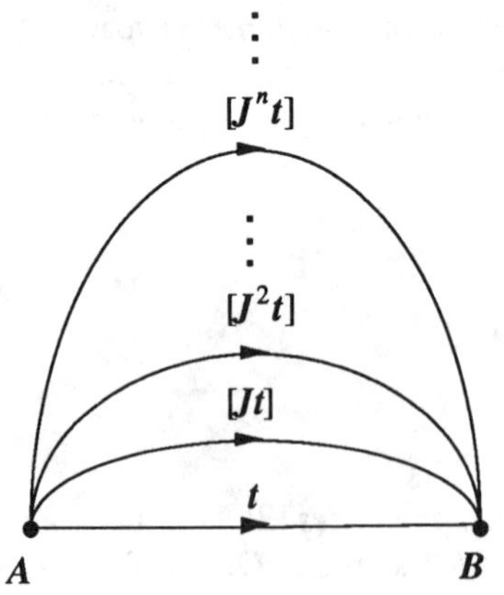

Figure 6.7 The expansion of the cyclic feedback branch shown in Figure 6.6.

We see that the cyclic feedback branch generates an infinite number of branches connected in parallel, with the nth order branch representing essentially n feedback branches connected in series. This picture suits well the intuitive conception of a feedback branch. If $\boldsymbol{J}$ is lossy (as it should generally be for the Neumann series to converge), then the power transmission of the various branches diminishes with n. Therefore, in practice we can truncate the graph in Figure 6.7 and retain only those branches with $n < N$, for a certain positive integer N. From this stage, the graph reduction may proceed using the other reduction rules.

Finally, we note that this treatment can be easily generalized for the case in which there is more then one incoming branch to the problematic node, or even to the case in which the incoming branches have cyclic transmissions themselves. We leave the consideration of these cases as an exercise for the reader.

6.4.2 Feedback Branches Containing Instantaneous Modulators

Let us consider the special case in which $\mathbf{J}(t_2,t_1)$ is in the form given in (6.12). It is not difficult to show that, in this case,

$$\mathbf{J}^n(t_2,t_1) = \mathrm{T}\{\prod_{k=0}^{n-1} \mathbf{V}(t_2 - k\tau)\}\,\delta(t_2 - t_1 - n\tau) \tag{6.38}$$

where $T\{\,\}$ stands for time ordering; namely, the various factors in the product have to be ordered in such a manner that their time arguments will increase from left to right. The order specification is necessary, since the matrices $\mathbf{V}(t_1)$ and $\mathbf{V}(t_2)$ may not commute for $t_1 \neq t_2$.

There are two interesting cases for which a relatively simple expression may be derived for $\mathbf{J}^n(t_2,t_1)$: the case of the frequency shifter and the case of the phase modulator. For a frequency shifter, we have

$$\mathbf{V}(t) = V_0\exp(-2\pi i f_0 t)$$

Using this relation in (6.38), we obtain

$$\mathbf{J}^n(t_2,t_1) = V_0{}^n\exp\{-\pi i n f_0[2t_2 - (n-1)\tau]\}\,\delta(t_2 - t_1 - n\tau) \tag{6.39}$$

The nth order branch transmission represents an equivalent frequency shifter with a frequency nf_0, as we could have expected. The Fourier expansion coefficients $(j^n)_k$ of $\mathbf{J}^n$ are computed from (6.13):

$$(j^n)_k = V_0{}^n\exp\{\pi i n\tau[2\nu + (n-1)f_0]\}\,\delta_{n,k} \tag{6.40}$$

For the phase modulator, we use

$$\mathbf{V}(t) = V_0\exp[-i\gamma\sin(2\pi f_0 t)] \tag{6.41}$$

where for simplicity we have dropped the constant phase ϕ, which is immaterial in this case since only one modulator is involved. Using this relation in (6.38), we obtain

$$\mathbf{J}^n(t_2,t_1) = V_0{}^n\exp\left\{-i\gamma\sum_{k=0}^{n-1}\sin[2\pi f_0(t_2 - k\tau)]\right\}\delta(t_2 - t_1 - n\tau) \tag{6.42}$$

The exponential in (6.42) is identical to the one that occurs in the transfer function of n instantaneous phase modulators successively shifted in phase by $-2\pi f_0\tau$. Substituting $-2\pi f_0\tau$ for $\Delta\phi$ in (6.25), we obtain

$$\mathbf{J}^n(t_2,t_1) = V_0^n \exp\{-i\gamma F_n(2\pi f_0\tau)\sin[\pi f_0(2t_2 - n\tau + \tau)]\}\delta(t_2 - t_1 - n\tau) \tag{6.43}$$

where we have used the fact that $F_n(x)$ is an even function of x. Thus, $\mathbf{J}^n(t_2,t_1)$ acts also as a phase modulator with a modified complex modulation index, but with an unmodified frequency. This is not surprising, since we have already found that a series combination of phase modulators is a phase modulator itself. The Fourier expansion coefficients $(j^n)_k$ of $\mathbf{J}^n(t_2,t_1)$ are found to be

$$(j^n)_k = V_0^n \exp\{i\pi\tau[2n\nu + k(n-1)f_0]\}J_k[\gamma F_n(2\pi f_0\tau)] \tag{6.44}$$

6.5 THE MODULATED RECIRCULATING LOOP

As an example of the feedback branch treatment in the time-dependent case, we consider the recirculating loop: with a modulator replacing the feedback waveguide. The physical scheme of this network is shown in Figure 6.8.

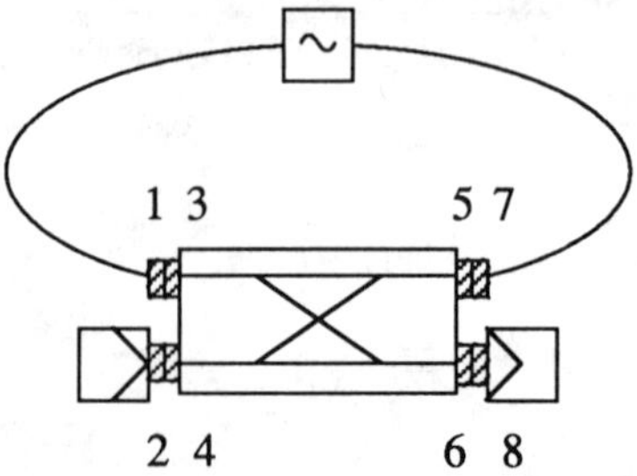

Figure 6.8 The physical scheme of a recirculating loop with a modulator replacing the feedback waveguide.

To compute the Fourier expansion coefficient set $\tilde{j}$ of the transfer matrix of this system, we quote the result that we obtained in Figure 5.5(d), replacing $(\boldsymbol{I} - t_{5,3}t_{3,5'})^{-1}$ by an infinite sum representing the Neumann series expansion of the feedback branch:

$$\tilde{j} = \tilde{t}_{4,6'} + [\tilde{t}_{3,6'} \sum_{n=0}^{\infty} [\tilde{t}_{5',3}\tilde{t}_{3,5'}]^n \tilde{t}_{5',3}\tilde{t}_{4,3}] \tag{6.45}$$

For the sake of simplicity, we assume that the Fourier expansion coefficient set $\tilde{t}_{5,3}$ of the modulator consists of scalar matrices of the form (see (6.13))

$$t_{5,3;k}(\nu) = e^{2\pi i\nu\tau} u_k(\nu) \boldsymbol{I} \tag{6.46}$$

Under this assumption, we can write

$$\tilde{j} = \tilde{t}_{4,6'} + [\tilde{t}_{3,6'} \sum_{n=1}^{\infty} \tilde{t}_{5',3}{}^n \tilde{t}_{3,5'}{}^{n-1} \tilde{t}_{4,3}] \tag{6.47}$$

To simplify the presentation even further, let us assume that all the transmissions in (6.47), with the exception of $\tilde{t}_{5',3}$ are independent of ν. Under this assumption, we can use the simplified rule (6.11a) for the multiplication of cyclic and time-invariant Fourier expansion coefficient sets to obtain:

$$j_k = t_{4,6'}\delta_{k,0} + t_{3,6'} \sum_{n=1}^{\infty} (t_{5',3}{}^n)_k t_{3,5'}{}^{n-1} t_{4,3} \tag{6.48}$$

Let us consider first the case in which the modulator is a frequency shifter; i.e.,

$$u_k = u_0 \delta_{k,1}$$

We can now use the result we derived earlier for the Fourier expansion coefficients of n frequency shifters combined in series (6.40) to get

$$j_k = t_{4,6'}\delta_{k,0} + t_{3,6'} \sum_{n=1}^{\infty} u_0{}^n \exp\{\pi i n\tau[2\nu + (n-1)f_0]\}\delta_{n,k} t_{3,5'}{}^{n-1} t_{4,3}$$

Performing the trivial summation, we get the final result:

$$j_k = t_{4,6'}\delta_{k,0} + u_0{}^k \exp\{\pi i k\tau[2\nu + (k-1)f_0]\} t_{3,6'} t_{3,5'}{}^{k-1} t_{4,5'} h_k \tag{6.49}$$

where $h_k = 1$ for $k > 0$ and $h_k = 0$ otherwise. We see that in this simple case we are able to obtain an expression for the Fourier expansion coefficients of the transfer function that does not involve an infinite sum in spite of the presence of the feedback branch. Also note that t_k is nonzero only for nonnegative k.

Let us assume that the coupler is both polarization- and power-preserving, and use for the various coupler transmissions the values that were used in Section 5.2. The system is now polarization-preserving as a whole, and the following expression is obtained for the Fourier expansion coefficient t_k of the transfer function of a single polarization mode:

$$j_k = e^{-i\phi}\cos\theta\{\delta_{k,0} - \mathrm{tg}^2\theta[u_0\exp[\pi i\tau(2n + kf_0 - f_0) + i\phi]\cos\theta]^k h_k\} \tag{6.50}$$

We turn now to the slightly more complicated case in which the time-periodic component is a phase modulator; i.e.,

$$u_k(\nu) = u_0 e^{2\pi i\nu\tau - ik\phi} J_k(\gamma) I$$

Using (6.44) we obtain

$$j_k = t_{4,6'}\delta_{k,0} + t_{3,6'}\sum_{n=1}^{\infty} u_0{}^n \exp\{i\pi\tau[2n\nu + k(n-1)f_0]\} J_k[\gamma F_n(2\pi f_0\tau)] t_{3,5'}{}^{n-1} t_{4,5'} \tag{6.51}$$

Let us invoke now the assumptions that led to (6.50) above. Doing this, we obtain the following expression for the kth Fourier expansion coefficient t_k of the transfer function:

$$j_k = e^{-i\phi}\cos\theta[\delta_{k,0} - \mathrm{tg}^2\theta Q_k(u_0\cos\theta e^{2\pi i\nu\tau - i\phi}, \gamma, f_0\tau)] \tag{6.52}$$

where

$$Q_k(\rho,\gamma,x) = \sum_{n=1}^{\infty} \rho^n e^{\pi ik(n-1)x} J_k[\gamma F_n(2\pi x)] \tag{6.53}$$

Considerable simplification occurs when the modulator is in resonance, that is, when $f_0\tau$ is an integer. Using (6.26) and the fact that [3]

$$J_n(-z) = (-1)^n J_n(z)$$

we obtain

$$Q_k(\rho,\gamma) = \sum_{n=1}^{\infty} \rho^n J_k(\gamma n) \tag{6.54}$$

The sum in (6.54) is a special case of the Schlömilch series [4]. Using the integral representation of the Bessel functions (Eq. (2.32)), we can transform it to an integral:

$$Q_k(\rho,\gamma) = \frac{1}{2\pi}\sum_{n=1}^{\infty}\rho^n \int_{-\pi}^{\pi} d\theta e^{-ik\theta + i\gamma n \sin\theta}$$

$$= \frac{\rho}{2\pi}\int_{-\pi}^{\pi} d\theta \, \frac{e^{-ik\theta + i\gamma \sin\theta}}{1 - \rho e^{i\gamma \sin\theta}} \tag{6.55}$$

This integral may be easier to evaluate than the sum in (6.54), since it involves elementary functions only.

6.6 THE COMBINATION OF TWO TRANSMISSIONS WITH COMMENSURATE FREQUENCIES

We conclude this chapter with a brief consideration of the problem of the combination of two cyclic transmissions with different periods. Our formalism can be applied to this case provided that the periods T_1 and T_2 involved are commensurate; i.e, there exist two integers N_1 and N_2 such that

$$N_1 T_1 = N_2 T_2 \tag{6.56}$$

In practice, of course, it can always be assumed that T_1 and T_2 are commensurate, since otherwise these periods must be specified with infinite precision. We define now a new period T:

$$T = N_1 T_1 = N_2 T_2 \tag{6.57}$$

and regard it as a period of both transmissions. We define now the Fourier expansion coefficients of both transmissions with respect to the new period T (which may result in a sparse set), and from this point we can apply the formalism of equal periods to combine the two transmissions in question.

6.7 SUMMARY

Networks containing modulators play an important role in many applications. Modulators are regarded here as components with a periodic time dependence. In general, a network may contain several modulators, each having a different frequency. If the network contains only one modulator, or if all modulators have the same frequency, then the network as a whole also has a periodic time dependence, and its frequency equals the common frequency of its modulators. This chapter has presented a general methodology for the analysis of such networks.

The analysis of time-periodic networks rests on the same fundamental ideas as the analysis of time-independent networks. The derivation of the signal flow graphs and the graphical reduction rules are identical in the time-independent and the time-periodic cases. The main difference lies in the frequency-domain product of the network algebra. While in the time-independent case this product is simply the algebraic matrix product, in the time-periodic case it becomes an infinite sum of such products. Infinite sums of matrices cannot generally be evaluated exactly; therefore, for the evaluation of the products of cyclic transmissions, it is necessary to resort to approximations, most commonly a truncation.

The advantage of using the frequency-domain representation over the time-domain representation is partially lost in the time-periodic case. Nevertheless, the frequency-domain representation is still preferable, since an infinite sum is generally easier to evaluate than an infinite integral. Another advantage of the frequency-domain approach is the fact that it leads directly to the evaluation of network Fourier expansion coefficients. As we will see, these quantities are necessary for the evaluation of the network effect on the optical signal. An exception to this assertion is the case of the instantaneous modulators. The time-domain product of the transmissions of such devices is reduced to the algebraic matrix product, while their frequency-domain product is still an infinite sum of matrix products. In the case of the instantaneous modulators, it is easier to evaluate the product in the time domain, and then evaluate the Fourier expansion coefficients of the final result, if required.

The method for evaluating the products of instantaneous modulators has been illustrated by analyzing the cases of amplitude modulators, frequency shifters, and phase modulators combined in series. We have shown that the serial combination of N frequency shifters or phase modulators results in an equivalent single-frequency shifter or phase modulator, respectively. Among other things, we show that it is possible to effectively increase the modulation index by combining two or more phase modulators in series. Special attention was given to the case of identical phase modulators combined in series and shifted successively by a constant phase. Such an arrangement is of practical interest, since it may describe the excitation of an ordered array of phase modulators by a travelling plane wave.

A more difficult problem is encountered when it becomes necessary to eliminate a cyclic feedback branch. This operation is equivalent to the solution of an integral equation in the time domain, and to the solution of a "sum" equation (the discrete analog of an integral equation) in the frequency domain. In the case in which the feedback branch represents a passive, lossy device, a solution using Neumann series expansion is generally possible.

The treatment of complete periodic networks was illustrated by analyzing the modulated Mach-Zehnder and recirculating loop systems. The latter also illustrates the treatment of a cyclic feedback branch.

Finally, the problem of combining two devices with different frequencies was considered. In practice, it is always possible to assume that the frequencies are commensurate; that is, it is possible to find a period (sometimes much higher than the fundamental) that is common to both devices. Once this period is found, it is possible to regard the components as having the same period and treat them with the formalism developed in this chapter.

References

[1] Born M. and E. Wolf, *Principles of Optics*, Oxford: Pergamon Press, Sixth Edition, 1980, pp. 401-407.

[2] Mathews, J. and R.L. Walker, *Mathematical Methods of Physics*, Redwood City: Addison-Wesley, Second Edition, 1970, pp. 302-305.

[3] Rainville, E.D., *Special Functions*, New York: Chelsea, 1960, pp. 109.

[4] Watson, G.N., *A Treatise on the Theory of Bessel Functions*, London: Cambridge University Press, Second Edition, 1944, pp. 618-653.

Chapter 7
Network Analysis of the Fiber-Optic Gyro

One of the most extensively studied optical networks in recent years is the fiber-optic gyro. The signal in the fiber-optic gyro is the Sagnac phase shift, which must be determined with a precision approaching 10^{-8} rad in the most demanding applications. Therefore, the analysis of high-performance fiber-optic gyro systems requires careful accounting for many physical phenomena that are normally neglected in ordinary optical networks. These phenomena include, among others, Rayleigh backscattering from the bulk of the fiber, mode coupling within the fiber, and the Kerr and the Faraday effects, all of which can give rise to spurious signals. The basic signal of the fiber-optic gyro is highly immune to environmental influences because of the reciprocity principle. However, this principle does not constrain the variations of the spurious signals, which can cause random fluctuations in the gyro functional parameters and degrade the overall performance. The elimination or control of the nonreciprocal spurious signals is the main goal of the high-performance fiber-optic gyro design.

The Sagnac phase shift appears in the transfer matrix of the gyro when it starts to rotate. In principle, the elements of the transfer matrix are measurable quantities. However, the measurement of these elements may require interferometric setups, since they are generally complex. For practical reasons, it is most desirable to reduce the measurement process to simple power monitoring with the help of an appropriate optical detector. Such a measurement setup can determine the matrix elements of the *power* transfer matrix only. Accordingly, we will regard here the power transfer matrix as the measurable quantity.

The modern fiber-optic gyro is a purely guided-wave, single-mode system, and as such is well suited for the analysis by the tools that we have developed. The analysis that we present here cannot be regarded as the state of the art, since the development of a detailed model accounting for all effects relevant to the gyro performance on the level required for inertial navigation purposes would span the volume of a separate book. Our analysis, however, will expose some of the main problems related to the fiber-optic gyro design, and explain the approaches for overcoming them. In addition, the present chapter serves as yet another, more sophisticated example of how to apply the optical network analysis in practice, highlighting polarization effects, which are crucial for the proper understanding of the fiber-optic gyro operation.

7.1 THE SAGNAC EFFECT IN THE ROTATING FIBER-OPTIC RING

In Figure 7.1 we show a two-port device consisting of a fiber coil wound on a drum with a radius R, similar in construction to the fiber-optic phase modulator that we have showed in Figure 2.3.

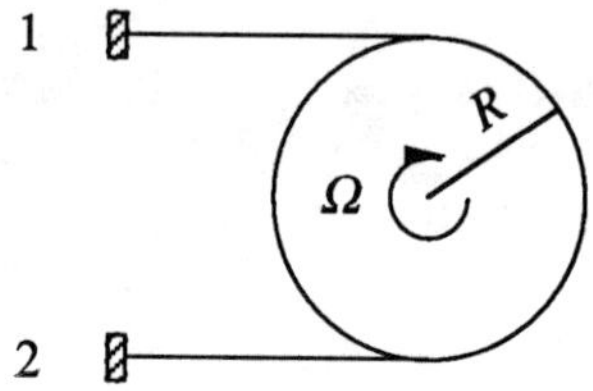

Figure 7.1 A fiber coil exhibiting the Sagnac effect.

When the fiber-optic coil is rotated about its axis with an angular velocity Ω, the effective time delays experienced by two waves propagating in it in opposing directions are no longer equal, and they differ by

$$2\tau_S = \frac{4\pi N R^2 \Omega}{c^2} \tag{7.1}$$

where N is the number of windings and c is the light velocity *in vacuum*. Alternatively, we may say that the two counterpropagating waves accumulate phase shifts that differ by

$$2\Phi = 2\pi\tau_S \nu = \frac{8\pi^2 N R^2 \Omega}{c^2} \nu \tag{7.2}$$

The angle 2Φ is called the *Sagnac phase shift*. According to (7.2), the Sagnac phase shift is proportional to the number of windings N, so it is possible to amplify the response of the device to rotation by increasing the fiber length in the coil. However the useful length is limited because of the fiber losses. In practical devices, fibers with a length of typically 1 km are used.

Taking into account the Sagnac effect, the signal flow graph of a rotating fiber coil is shown in Figure 7.2, where $\boldsymbol{F}$ and $\boldsymbol{F}^T$ denote the corresponding Jones matrices of the coil at rest (i.e., with $\Omega = 0$). The Sagnac phase shift 2Φ as defined in Figure 7.2 is positive when the coil rotates in the clockwise direction. By taking the rest Jones matrix $\boldsymbol{F}^T$ of the lower arm to be the transpose of the corresponding Jones matrix $\boldsymbol{F}$ of the upper arm, we implicitly assumed that the coil is intrinsically reciprocal, as is commonly the case. However, when the coil starts to rotate, the reciprocity is destroyed, and the lower arm Jones matrix is no longer the transpose of the upper arm Jones matrix, as indicated explicitly in Figure 7.2. Since reciprocity is satisfied to a high degree by the other network components, its violation by the rotating coil is highly conspicuous, thus allowing a very sensitive detection and measurement of Ω. The operation principle of the fiber-optic gyro is based upon this observation.

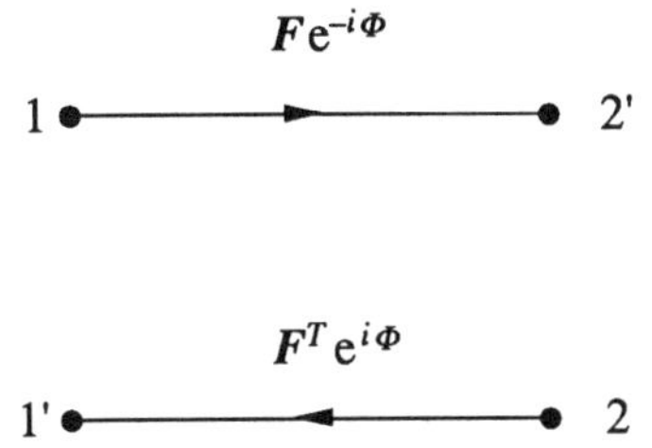

Figure 7.2 The signal flow graph of a rotating fiber coil.

For a circular path in a vacuum, the Sagnac phase shift may be calculated by classical time-of-flight considerations, accounting for the rotation of the path during the transit time [1]. This approach fails when applied to the fiber-optic coil, since it would result in formulas analogous to (7.1) and (7.2), with c replaced by the velocity of light in the fiber. The correct expression for the Sagnac phase shift is regained only when the time-of-flight argument is augmented by the Fresnel-Fizeau drag effect [1]. A rigorous treatment of the Sagnac effect is possible within the framework of the relativistic formulation of electrodynamics [1].

In all of our examples considered in Chapter 5, we assumed that the waveguide Jones matrix $\boldsymbol{F}$ is scalar; i.e.,

$$\boldsymbol{F} = e^{2\pi i\nu\tau}\boldsymbol{I} \tag{7.3}$$

where τ is the waveguide delay time. This assumption is not adequate for the modelling of the fiber-optic gyro. Even if the original fiber is perfectly degenerate and polarization-preserving, its mere winding on the drum introduces both birefringence and mode coupling, which can degrade the gyro performance to a level that may exclude its application for inertial navigation purposes. We will assume, however, that $\boldsymbol{F}$ is independent of Φ; that is, it is not modified by the rotation of the coil. Even though it is possible to think of physical mechanisms that will make $\boldsymbol{F}$ dependent on Φ (such as the generation of extra birefringence due to the centrifugal force), to the best of our knowledge such effects have not been discussed in the literature, and they can probably be neglected in the cases of interest.

In spite of the fact that we allow $\boldsymbol{F}$ to have a general form, Figure 7.2 is not the most general representation of a fiber coil, since we have neglected reflections coming from the fiber bulk, which are usually dominated by the Rayleigh scattering mechanism [2,3]. To account for these reflections, we should have included in the signal flow graph of Figure 7.2 two vertical branches connecting the upper and the lower nodes. The consideration of such reflections is essential for the evaluation of the gyro performance [4]. We will not consider these effects here.

To get an idea of the precision required for the measurement of the Sagnac phase, let us evaluate (7.2) for a realistic case. For inertial navigation applications, a gyro must be able to measure absolute angular velocities with a precision of 10^{-3} deg/h (which is four orders of magnitude slower than the earth's angular velocity). For this angular velocity and some typical fiber-optic gyro parameters, namely, $R = 0.1\text{m}$, $N = 10^4$ and $\nu = 3 \times 10^{14}$ Hz (corresponding to the near infrared), we get $2\Phi \approx 2 \times 10^{-8}$ rad. (Note that for the evaluation of (7.2), Ω must first be converted to units of rad/sec.) The total phase acquired by a wave propagating in a 1-km-long fiber is on the order of 10^{10}; therefore, for inertial navigation purposes, the phase must be determined with a relative precision of 1 part in 10^{18}.

7.2 A BASIC FIBER-OPTIC SAGNAC INTERFEROMETER

In order to expose the problems involved in fiber-optic gyro design, we consider first a simple system that is, in principle, capable of measuring the Sagnac phase shift. The physical scheme and the signal flow graph of such a system is shown in Figure 7.3. The interference of the counterpropagating waves in this system was demonstrated first by Vali and Shorthill [5]. In the original experiment, a bulk-optics beam splitter was used instead of the coupler.

The transfer matrix $\boldsymbol{J}$ of the system shown in Figure 7.3 is given by

$$\boldsymbol{J} = \boldsymbol{t}_{3,2'}\boldsymbol{F}\boldsymbol{t}_{1,4'}e^{i\Phi} + \boldsymbol{t}_{4,2'}\boldsymbol{F}^T\boldsymbol{t}_{1,3'}e^{-i\Phi} \tag{7.4}$$

As a zero-order approximation, let us assume that the coupler is ideally reciprocal and power- and polarization-preserving, and that $\boldsymbol{F}$ is given by (7.3). Under these assumptions, the system is polarization-preserving as a whole, and we may consider the scalar version of (7.4) which applies to a single polarization mode:

$$J = e^{2\pi i\nu\tau}(t_{3,2'}t_{1,4'}e^{i\Phi} + t_{4,2'}t_{1,3'}e^{-i\Phi}) \tag{7.5}$$

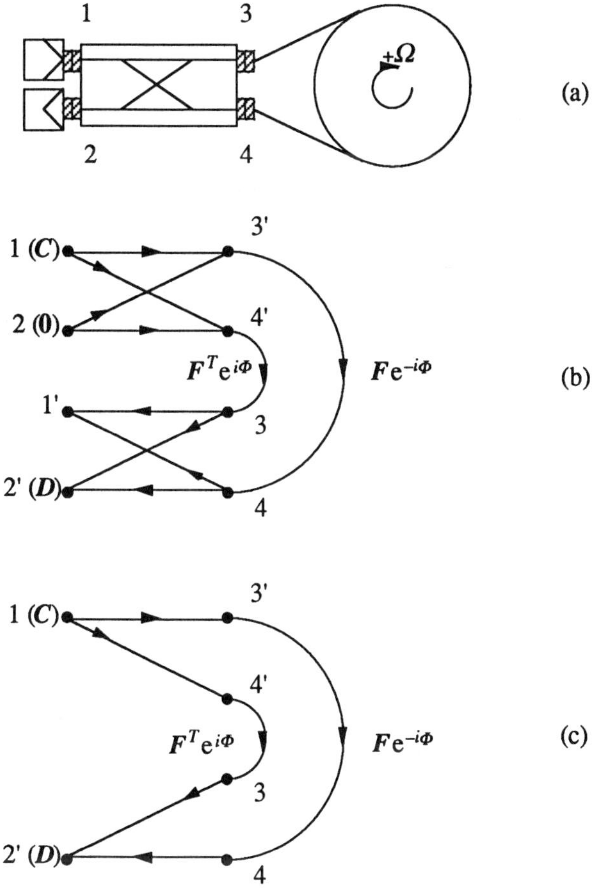

Figure 7.3 A simple Sagnac interferometer: (a) the physical scheme; (b) the signal flow graph; (c) the signal flow graph without the irrelevant nodes.

We further use the coupler parametrization (Eq. (5.6)) to assign the following values for the various coupler transmissions:

$$t_{3,2'} = e^{i\psi}\sin\theta, \qquad t_{4,2'} = e^{-i\phi}\cos\theta$$

$$t_{1,4'} = e^{i(\pi-\psi)}\sin\theta, \qquad t_{1,3'} = e^{i\phi}\cos\theta$$

Using these values in (7.5), we obtain

$$J = e^{2\pi i\nu\tau}(\cos^2\theta e^{-i\Phi} - \sin^2\theta e^{i\Phi}) \tag{7.6}$$

Since the detector responds to the optical power only, the Sagnac phase shift is accessible for measurement only through the power transfer function $|J|^2$:

$$|J|^2 = \cos^4\theta + \sin^4\theta - 2\sin^2\theta\cos^2\theta\cos(2\Phi) \tag{7.7}$$

which depends explicitly on Φ, thus allowing, in principle, its measurement.

The simple Sagnac interferometer has two obvious shortcomings concerning the measurement of Φ. First, the sign of Φ cannot be determined; second, for small rotation rates the system response varies as Ω^2, significantly degrading the sensitivity. Both these shortcomings can be eliminated by the introduction of a proper biasing mechanism, as we will discuss later. However, the simple Sagnac interferometer suffers from more fundamental deficiencies.

7.3 PROBLEMS ARISING FROM BIREFRINGENCE AND MODE MIXING

In arriving at (7.7), we have assumed that all the components have an ideal behavior. In order to assess the performance of the simple Sagnac interferometer as a practical gyro, we have to find out how deviations from the ideal behavior affect the device. Let us start with the fiber coil.

As we already mentioned, the mere winding of the fiber on the drum introduces birefringence. To account for birefringence, let us assume that $\boldsymbol{F}$ has the form

$$\boldsymbol{F} = e^{2\pi i\nu\tau}\boldsymbol{B}(\beta L) \tag{7.8}$$

where L is the total fiber length in the coil, β the specific birefringence, and

$$\boldsymbol{B}(x) = \begin{pmatrix} e^{ix} & 0 \\ 0 & e^{-ix} \end{pmatrix} \tag{7.9}$$

There are two main contributions to the specific birefringence β: one contribution comes from the intrinsic composition of the fiber itself, and the other is due to the stresses induced in it by the winding. The contribution due to the winding alone amounts to approximately 500 deg/m [6]; the magnitude of the other one depends on the quality of the fiber, but is typically comparable in magnitude to the winding contribution. Thus, the phase shifts induced by birefringence effects are many orders of magnitude greater than the Sagnac phase shift, and obviously cannot be disregarded. However, birefringence by itself does not degrade the performance of the simple Sagnac interferometer as a gyro; the degradation takes place because of the combination of birefringence and the presence of mode mixing in the coil [7].

In order to understand how the degradation occurs, let us consider a rather extreme (and artificial) case of a *mode switcher* being inserted between the fiber coil and one of the coupler ports, as shown in Figure 7.4.

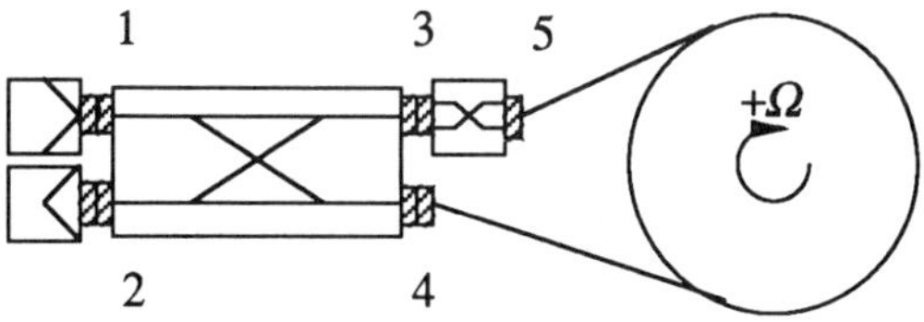

Figure 7.4 The simple Sagnac interferometer with an interfering mode switcher.

The mode switcher is a directional two-port device with a transmission X given by

$$X = \begin{pmatrix} 0 & 1 \\ 1 & 0 \end{pmatrix} \tag{7.10}$$

A device with such a transmission switches the fields between the two modes; i.e.,

$$X \begin{pmatrix} a \\ b \end{pmatrix} = \begin{pmatrix} b \\ a \end{pmatrix}$$

Let us consider now the phase shift between the two counterpropagating field components reaching the detector in mode 1. The field component that propagated through the coil in the clockwise direction acquired in the coil a total phase ϕ_{CW} of

$$\phi_{CW} = 2\pi\nu\tau + \beta L + \Phi$$

since it had to propagate through the coil in mode 1 in order to reach the detector in that mode. On the other hand, the field component that propagated through the coil in the counterclockwise direction acquired in the coil a total phase ϕ_{CCW} of

$$\phi_{CCW} = 2\pi\nu\tau - \beta L - \Phi$$

since it had to propagate through the coil in mode 2 in order to reach the detector in mode 1 due to the action of the mode switcher. Consequently, the phase shift $\Delta\phi$ between the two counterpropagating field components reaching the detector in mode 1 is

$$\Delta\phi = \phi_{CW} - \phi_{CCW} = 2(\beta L + \Phi)$$

Similar considerations regarding the field component reaching the detector in mode 2 yield a phase shift of $2(\beta L - \Phi)$.

Let us see now how these conclusion can be reached formally using network analysis. The signal flow graph corresponding to Figure 7.4 is shown in Figure 7.5. The transfer matrix $\boldsymbol{J}$ of this system is given by

$$\boldsymbol{J} = e^{2\pi i\nu\tau}(\cos^2\theta \boldsymbol{B}\boldsymbol{X}e^{-i\Phi} - \sin^2\theta \boldsymbol{X}\boldsymbol{B}e^{i\Phi}) \tag{7.11}$$

From this we compute the power transfer matrix $\lfloor \boldsymbol{J} \rfloor^2$:

$$\lfloor \boldsymbol{J} \rfloor^2 = (\cos^4\theta + \sin^4\theta)\boldsymbol{I} - 2\sin^2\theta\cos^2\theta \begin{pmatrix} \cos[2(\beta L + \Phi)] & 0 \\ 0 & \cos[2(\beta L - \Phi)] \end{pmatrix} \tag{7.12}$$

Thus, $\lfloor \boldsymbol{J} \rfloor^2$ depends on Φ only through the sum $\beta L + \Phi$ and the difference $\beta L - \Phi$, as we have concluded earlier by using heuristic arguments.

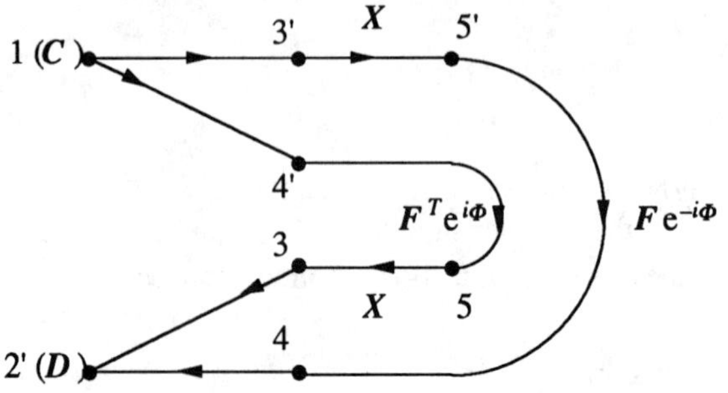

Figure 7.5 The signal flow graph of the simple Sagnac interferometer with the mode switcher.

At first sight it may seem that the presence of the birefringence and the mode switcher does not hinder the measurement of Φ: all that has to be done is to measure θ and the rotation-independent phase shift $2\beta L$, an operation which is essentially a calibration procedure. Once these parameters are known, the Sagnac phase shift can be measured by monitoring the optical power incident on the detector. This procedure, however, fails in practice. The reason for this failure is the fact that, in practice, β inevitably contains a randomly fluctuating component, due to the influence of environmental conditions such as temperature and vibrations. As we have seen, for inertial navigation purposes, the Sagnac phase shift has to be determined with a precision of 10^{-8} rad. This means that the fluctuations in β for a 1-km fiber must be suppressed to a level of 10^{-11} rad/m. Since the typical value of β is 5rad/m, this means that the relative suppression of the fluctuations must be on the level of one part in 5×10^{11}. Such a degree of isolation from the environment is very difficult to achieve in practice.

Mode mixing occurs within the fiber because of intrinsic structural imperfections and external perturbations. Though it is highly unlikely that a complete mode switching like the one that we have considered above will occur within the fiber, a certain amount of mode mixing is practically impossible to prevent. This means that in order to analyze the fiber-optic gyro we are not allowed to neglect the nondiagonal elements in $\boldsymbol{F}$, and, in particular, we cannot use the approximation in (7.3). Using in (7.4) the transmissions of an ideal coupler but retaining a general $\boldsymbol{F}$ we obtain

$$\boldsymbol{J} = \cos^2\theta \boldsymbol{F}^T e^{-i\Phi} - \sin^2\theta \boldsymbol{F} e^{i\Phi} \tag{7.13}$$

The corresponding power transfer matrix is

$$|\boldsymbol{J}|^2 = \cos^4\theta \boldsymbol{F}^\dagger \boldsymbol{F} + \sin^4\theta \boldsymbol{F}^* \boldsymbol{F}^T - \sin^2\theta\cos^2\theta(\boldsymbol{F}^\dagger \boldsymbol{F}^T e^{i\Phi} + \boldsymbol{F}^* \boldsymbol{F} e^{-i\Phi}) \tag{7.14}$$

The matrix $\boldsymbol{F}^\dagger \boldsymbol{F}^T$ (as well as its hermitian conjugate $\boldsymbol{F}^*\boldsymbol{F}$), will contain, in practice, phases with random components. These phases will add up to Φ, thus obscuring its measurement, as explained above.

The problems that arise from random fluctuations in the phases of $\boldsymbol{F}$ can be eliminated by adding two polarizers: one between the coupler and the source and another between the coupler and the detector, as shown in Figure 7.6(a) [7]. Referring to Figure 7.6(b), the transfer matrix $\boldsymbol{J}$ of this system is given by

$$\boldsymbol{J} = \cos^2\theta \boldsymbol{P}\boldsymbol{F}^T\boldsymbol{P} e^{-i\Phi} - \sin^2\theta \boldsymbol{P}\boldsymbol{F}\boldsymbol{P} e^{i\Phi}$$

where $\boldsymbol{P}$ is the Jones matrix of the polarizers, which we assume to be of the form

$$\boldsymbol{P} = \begin{pmatrix} 1 & 0 \\ 0 & 0 \end{pmatrix}$$

The transfer matrix $\boldsymbol{J}$ can be also cast in the form

$$\boldsymbol{J} = F_{11}(\cos^2\theta e^{-i\Phi} - \sin^2\theta e^{i\Phi})\boldsymbol{P} \tag{7.15}$$

since for any matrix $\boldsymbol{A}$,

$$\boldsymbol{PAP} = A_{11}\boldsymbol{P} \tag{7.16}$$

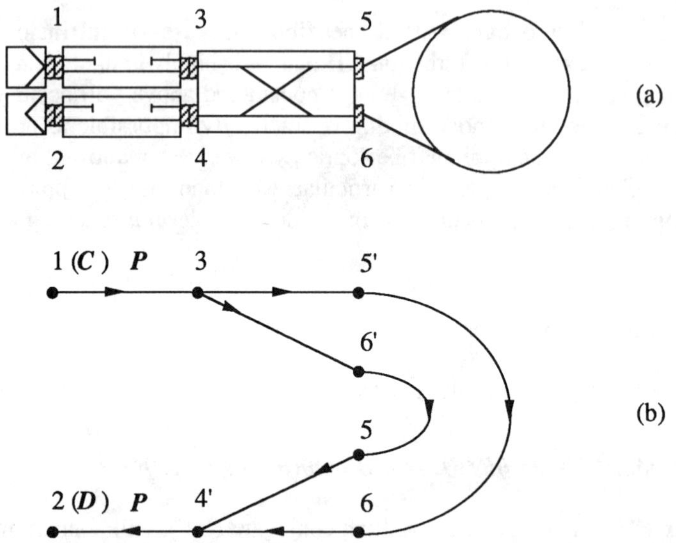

Figure 7.6 An improved Sagnac interferometer with polarizers: (a) The physical scheme; (b) the signal flow graph.

Thus, the power transfer matrix for the improved Sagnac interferometer is

$$|\boldsymbol{J}|^2 = |F_{11}|^2[\cos^4\theta + \sin^4\theta - 2\sin^2\theta\cos^2\theta\cos(2\Phi)]\boldsymbol{P} \tag{7.17}$$

in which all randomly varying phases of $\boldsymbol{F}$ have been eliminated. It is easy to understand why this elimination occurs: the two polarizers ensure that only those field components that propagated in the coil in mode 1 will reach the detector. By reciprocity, fields that propagate through the coil in the same mode acquire at rest (in the absence of magnetic fields) the same phase, regardless of the direction of the propagation [8]. Thus, in this situation, the only phase shift that can arise between the two counterpropagating waves must be due to the Sagnac effect.

7.4 PROBLEMS ARISING FROM COUPLER LOSSES

A more serious problem with the simple Sagnac interferometer arises from the fact that real couplers cannot be ideally power-preserving. The relaxation of the unitarity requirement from the S-matrix parametrization of the coupler introduces new free parameters, some of which appear as phases that will add up to the Sagnac phase shift. To see how this may come about, let us assume that the S-submatrix s_c of the coupler is of the form

$$s_c = \begin{pmatrix} 0 & 0 & re^{i\phi} & se^{i(\pi-\psi+\varepsilon)} \\ 0 & 0 & se^{i\psi} & re^{-i\phi} \\ re^{i\phi} & se^{i\psi} & 0 & 0 \\ se^{i(\pi-\psi+\varepsilon)} & re^{-i\phi} & 0 & 0 \end{pmatrix} \tag{7.18}$$

where s and r are two real constants such that

$$s^2 + r^2 < 1 \tag{7.19}$$

This S-submatrix describes a lossy, reciprocal, and polarization-preserving coupler. For such a coupler we have

$$t_{3,2'} = se^{i\psi}, \qquad t_{4,2'} = re^{-i\phi}$$

$$t_{1,4'} = se^{i(\pi-\psi+\varepsilon)}, \qquad t_{1,3'} = re^{i\phi}$$

Using these values for the transmissions in (7.5), we obtain

$$J = e^{2\pi i\nu\tau}[r^2e^{-i\Phi} - s^2e^{i(\Phi+\varepsilon)}]$$

and the power transfer function becomes

$$|J|^2 = r^4 + s^4 - 2r^2s^2\cos(2\Phi + \varepsilon) \tag{7.20}$$

The quantity ε represents an accumulated phase shift, and is on the order of the size of the coupler divided by the wavelength. As in the case of the birefringence phase shift, ε is extremely sensitive to environmental influence. This is due to the fact that the physical size of the coupler is typically 10^4 wavelengths, which means that any change in the index of refraction n of the coupler material is multiplied by 10^4 in ε. Controlling ε on the level of 10^{-8} rad requires, therefore, the suppression of fluctuations in n on the level of 10^{-12}, which is extremely difficult in practice; a change of only 1 °C modifies the index of refraction of silica by 10^{-5} [9]. This problem obviously cannot be remedied by the use of polarizers like the previous one, since it is not generated by polarization effects. It is actually due to the fact that in the simple Sagnac interferometer the paths of the two counterpropagating fields in the coupler are not equivalent, thus allowing for the acquisition of spurious phases that are not constrained by reciprocity and therefore do not cancel out.

7.5 A PRACTICAL FIBER-OPTIC GYRO

The problems arising from the imperfections in the coupler can be eliminated by using the design shown in Figure 7.7(a) [10].

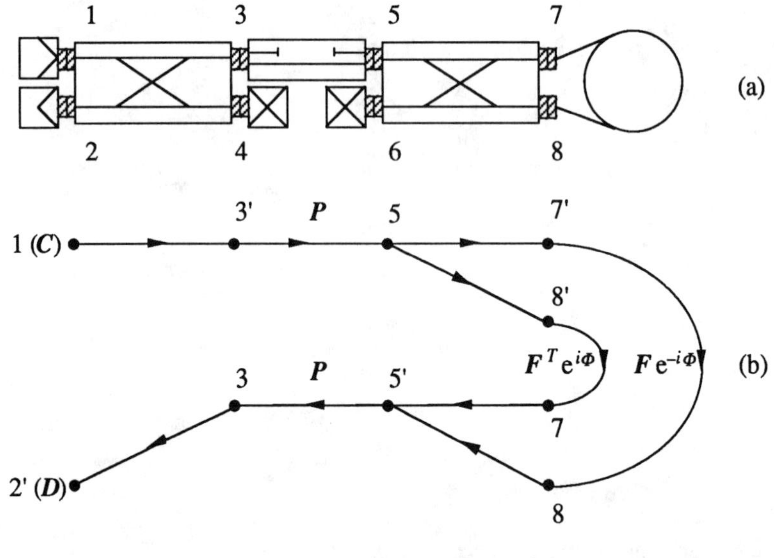

Figure 7.7 A practical fiber-optic gyro [10]: (a) the physical scheme; (b) the signal flow graph.

In this configuration, as in the simple Sagnac interferometer, the Sagnac phase shift is generated by a fiber coil connected to a coupler. However, this time the interfering fields are sampled at the same port that is used for the injection, namely, port number

5 in Figure 7.7(a); whereas in the simple Sagnac interferometer different ports are used for these purposes. This architecture causes both counterpropagating fields to acquire the same phases when passing through the coil coupler, which subsequently are cancelled out. To see how this comes about formally, let us compute the transfer matrix $\boldsymbol{J}$ of this setup. Referring to Figure 7.7(b), we get

$$\boldsymbol{J} = t_{3,2'}\boldsymbol{P}(e^{i\Phi}t_{7,5'}\boldsymbol{F}^T t_{5,8'} + e^{-i\Phi}t_{5',8}\boldsymbol{F}t_{5,7'})\boldsymbol{P}t_{1,3'}$$

By reciprocity we get

$$t_{7,5'} = t_{5,7'}{}^T$$

and

$$t_{5,8'} = t_{5',8}{}^T$$

We can therefore cast $\boldsymbol{J}$ in the form

$$\boldsymbol{J} = t_{3,2'}\boldsymbol{P}(e^{i\Phi}\boldsymbol{A}^T + e^{-i\Phi}\boldsymbol{A})\boldsymbol{P}t_{1,3'} \tag{7.21}$$

where

$$\boldsymbol{A} = t_{5',8}\boldsymbol{F}t_{5,7'}$$

By virtue of (7.16) we have

$$\boldsymbol{J} = 2A_{11}\cos\Phi\, t_{3,2'}\boldsymbol{P}t_{1,3'} \tag{7.22}$$

so that the power transfer matrix $|\boldsymbol{J}|^2$ is given by

$$|\boldsymbol{J}|^2 = 4|A_{11}|^2\cos^2\Phi|(t_{3,2'})_{11}|^2 t^\dagger{}_{1,3'}\boldsymbol{P}t_{1,3'} \tag{7.23}$$

If we use for $t_{1,3'}$ the value $re^{i\phi}\boldsymbol{I}$, we obtain

$$|\boldsymbol{J}|^2 = 4r^2|A_{11}|^2\cos^2\Phi|(t_{3,2'})_{11}|^2\boldsymbol{P} \tag{7.24}$$

In general, the transmission $t_{1,3'}$ may be nondiagonal (which would imply that the coupler is not polarization-preserving). In this case, the nondiagonal elements of the power transfer matrix may still depend on some fluctuating phases originating from the nondiagonal elements of $t_{1,3'}$. Later we will see, however, that if the two polarization modes of the source are uncorrelated (in a suitably chosen representation), then the detector output will be independent of the nondiagonal elements of the power transfer matrix. The sources that are used for fiber-optic gyro devices (laser or superluminescent diodes) are normally intrinsically planarly polarized; the weak orthogonal polarization mode, as far as we know, is uncorrelated with the principal mode. The dependence of the detector output on the nondiagonal elements of the power transfer matrix can be eliminated with such sources, provided that the polarizer axis is aligned with the principal polarization mode of the source. This alignment can be achieved using a device known as the *polarization controller*, which has the capability to rotate the polarization plane of the field guided within a degenerate fiber by an arbitrary angle. In addition to its effect on the stability of the system, such an alignment also optimizes the optical throughput.

To sum up, the architecture shown in Figure 7.7 allows a very accurate measurement of the rotation rate Ω, in a way which is highly immune to environmental influence, thus offering a long-term stability. These properties are maintained regardless of the presence of a possible mode coupling and birefringence in the coil fiber, and losses in the coupler, as a direct consequence of the reciprocity principle. We should mention, though, that this principle applies to strictly time-independent systems only, and time-dependent perturbations with a spectrum overlapping the detection bandpass will generate noise in the gyro output [9].

7.6 THE INTRODUCTION OF BIAS BY PHASE MODULATION

Both the one- and the two-coupler architectures suffer from the fact that their differential response to Ω vanishes at rest, since their power transfer matrix depends on the Sagnac phase shift Φ through $\cos\Phi$. In order to allow accurate measurement of small rotation rates, it is necessary to modify the dependence of the power transfer matrix on Ω so as to make it linear for small rotation rates. This is achieved by an appropriate shift in the rest operating point of the gyro, or, in other words, by the introduction of bias.

Several mechanisms have been proposed for the generation of bias, but the most successful one is the introduction of a phase modulator at one of the ends of the coil [11], as shown in Figure 7.8. For simplicity we assume that the modulator is a component with an instantaneous time response, so that its Fourier expansion coefficients $\boldsymbol{m}_k$ and $\boldsymbol{m'}_k$ are independent of ν. We also assume, as usual, that the various coupler transmissions are independent of ν as well.

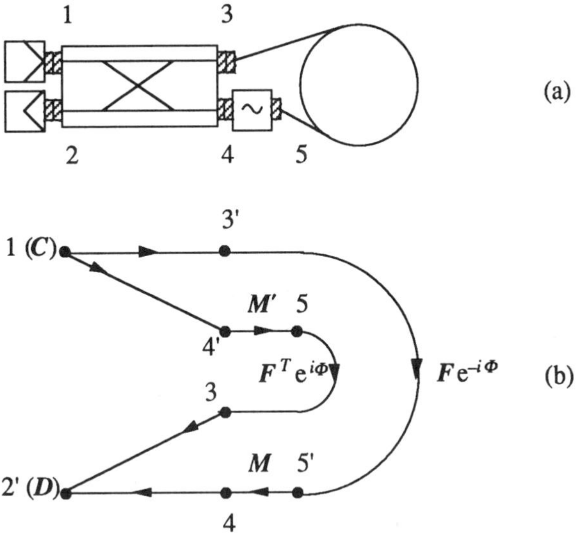

Figure 7.8 A simple Sagnac interferometer with a modulator for the generation of bias: (a) the physical scheme; (b) the signal flow graph.

Using the rules for the product of cyclic and time-independent transmissions (Eqs. (6.10) and (6.11)), we get the following expression for the kth Fourier expansion coefficient j_k of the transfer matrix:

$$j_k = t_{4,2'}m_k\exp[2\pi i\nu(\tau - \tau_S)]t_{1,3'} + t_{3,2'}\exp[2\pi i(\nu + kf_0)(\tau + \tau_S)]m'_k t_{1,4'}$$

Neglecting $f_0\tau_S$ and rearranging, we get

$$\begin{aligned} j_k &= \exp[\pi i\tau(2\nu + kf_0)]\{t_{4,2'}m_k t_{1,3'}\exp[-\pi i(2\nu\tau_S + kf_0\tau)] \\ &\quad + t_{3,2}m'_k t_{1,4'}\exp[\pi i(2\nu\tau_S + kf_0\tau)]\} \end{aligned} \quad (7.25)$$

According to the discussion following (2.24), if the source is monochromatic with a frequency ν_0 then the optical power in the output field component at the frequency $\nu_0 + kf_0$ is proportional to $|j_k|^2$. To see that the dependence of this optical power on Φ is biased, let us assume again that the system is polarization-preserving, and replace the various transmissions with their values for a power-preserving coupler. Furthermore, we assume that $m_k = m'_k$, which is a plausible assumption for a polarization-preserving modulator. In these circumstances, for a single polarization mode, we obtain

$$j_k = \exp[\pi it(2\nu + kf_0)]m_k\{\cos^2\theta\exp[-\pi i(2\nu\tau_S + kf_0\tau)]$$
$$-\sin^2\theta\exp[\pi i(2\nu\tau_S + kf_0\tau)]\} \tag{7.26}$$

and

$$|j_k|^2 = |m_k|^2[\cos^4\theta + \sin^4\theta - 2\sin^2\theta\cos^2\theta\cos(2\Phi + \pi kf_0\tau)] \tag{7.27}$$

Thus, for instance, by taking $f_0\tau = 1/2$, we obtain maximal differential sensitivity at $\Phi = 0$ for odd k.

To take advantage of this scheme, it is necessary to monitor the optical power at one of the shifted frequency components of the output field. Although this may be accomplished with an appropriate optical filter, it is usually easier to carry out the filtration electronically on the output current of the detector. The general methods for the analysis of the detector output will be described in detail in Part 2 of this book. We postpone the discussion of the detector output of the modulated fiber-optic gyro until the end of Chapter 12.

7.7 SUMMARY

The fiber-optic gyro has attracted considerable interest in the past few years because of its potential applicability as the rotation sensor of inertial navigation platforms. Modern fiber-optic gyro designs use exclusively guided-optics components, so that the system as a whole may be regarded as a guided optical network. The basic fiber-optic gyro architecture contains five optical components: two directional couplers, polarizer, polarization controller, and phase modulator.

The operation principle of the fiber-optic gyro is based on the Sagnac effect. This effect is manifested by the appearance of a phase difference between the two counterpropagating fields in a rotating fiber coil. This phase difference is called the Sagnac phase shift, and it is proportional to the angular velocity of the coil. Thus the measurement of the Sagnac phase shift allows the determination of the rotation rate of the coil.

For inertial navigation applications, the Sagnac phase shift must be measured with a precision of 10^{-8} rad, which is equivalent to a precision of 10^{-3} deg/h in the rotation rate. This high precision puts very stringent requirements on the stability of the network power transfer matrix. The stabilization of the relevant scattering parameters that contribute to the network power transfer matrix is the main design goal of the fiber-optic gyro system. The phases of the scattering parameters are normally most susceptible to external perturbations, and are therefore most difficult to stabilize. In practical devices these phases must be eliminated from the power transfer matrix to allow a stable operation. Such a cancellation of the uncontrolled phases can be achieved with the two-coupler architecture.

The fiber-optic gyro is an interferometer. Like any other interferometric structure, the dependence of its transfer matrix on the phase shift is periodic. The basic Sagnac interferometer has a zero sensitivity at rest. In a practical device it is necessary to introduce a constant phase shift that will move the operating point of the interferometer to the point of maximal sensitivity at rest. This constant phase shift is called bias. Various schemes for the introduction of bias in fiber-optic gyros were proposed. The most popular scheme is based on the introduction of a phase modulator close to one of the edges of the interferometer coil. In this scheme, the intensities of the shifted output field components are biased, and a biased operation can be achieved by monitoring the intensity of one of them with an appropriate optical filter. In practice, it is easier to recover the biased signal by filtering electronically the detector output. This method will be discussed in Chapter 12.

References

[1] Post, E.J. "Sagnac Effect," *Rev. Mod. Phys.*, Vol. 39, No. 2, 1967, pp. 475-493.

[2] Hartog A.H. and M.P. Gold, "On the Theory of Backscattering in Single-Mode Optical Fibers," *IEEE J. Lightwave Technol.*, Vol. LT-2, 1984, p. 76.

[3] Lin S-C. and T.G. Giallorenzi, "Sensitivity Analysis of the Sagnac-Effect Optical-Fiber Ring Interferometer," *Appl. Opt.*, Vol. 18, No. 6, 1979, pp. 915-931.

[4] Cutler, C.C., S.A. Newton, and H.J. Shaw, "Limitation of Rotation Sensing by Scattering," *Opt. Lett.*, Vol. 5, No. 11, 1980 pp. 488-490.

[5] Vali V. and R.W. Shorthill, "Fiber Ring Interferometer," *Appl. Opt.*, Vol. 15, No. 5, 1976, pp. 1099-1100.

[6] Ulrich R., S.C. Rashleigh and W. Eickhoff, "Bending-Induced Birefringence in Single-Mode Fibers," *Opt. Lett.*, Vol. 5, No. 6, 1980, pp. 273-275.

[7] Ulrich R. and M. Johnson, "Fiber-Ring Interferometer: Polarization Analysis," *Opt. Lett.*, Vol. 4, No. 5, 1979, pp. 152-154.

[8] Schiffner G., W.R. Leeb, H. Krammer, and J. Wittmann, "Reciprocity of Birefringent Single-Mode Fibers for Optical Gyros," *Appl. Opt.*, Vol. 18, No. 13, 1979, pp. 2096-2097.

[9] Shupe D.M., "Thermally Induced Nonreciprocity in the Fiber-Optic Interferometer," *Appl. Opt.*, Vol. 19, No. 5, 1980, pp. 654-655.

[10] Bergh R.A., H.C. Lefevre, and H.J. Shaw, "All-Single-Mode Fiber-Optic Gyroscope with Long-Term Stability," *Opt. Lett.*, Vol. 6, No. 10, 1981, pp. 502-504.

[11] Ulrich, R., "Fiber-Optic Rotation Sensing with Low Drift", *Opt. Lett.*, Vol. 5, No. 5, 1980, pp. 173-175.

[12] Abramowitz M. and I.A. Stegun, *Handbook of Mathematical Functions*, New York: Dover Publications, 1968, p. 361.

Further Reading

Bergh, R.A., H.C. Lefèvre, and H.J. Shaw, "An Overview of Fiber-Optic Gyroscopes," *IEEE J. Lightwave Technol.*, Vol. LT-2, No. 2, 1984, pp. 91-107.

Lefèvre, H.C., "Fiber-Optic Gyroscope," in *Optical Fiber Sensors: Systems and Applications*, B. Culshaw and J. Dakin, eds., Norwood: Artech House, 1989, pp. 381-427.

Part 2:

Signal Analysis

Chapter 8
The Second-Order Statistics of Guided Fields

The generation of the optical field by continuous-wave, solid-state optical sources used in optical networks is typically a narrowband and stationary stochastic process. The spectral purity of these sources varies over many orders of magnitude: it may be as narrow as 100 KHz in highly coherent laser diodes and as large as tens of THz for light-emitting diodes. Accordingly, different statistical models are used to describe the properties of the various optical sources. The generation of the optical field by highly coherent, amplitude-stabilized lasers is modelled as a random-phase process, and light-emitting diodes are regarded as thermal sources that can be described by Gaussian statistics. More complex statistical models may be required to describe optical sources that lie somewhere in between these two extremes.

In general, arbitrary correlations may exist between the polarization modes of the field. In practice, however, we usually deal with fields in which the two polarization modes are statistically independent. We have called such fields separable fields. The analysis of separable fields is considerably simpler than the analysis of general fields. The fundamental results will be presented for the general case; however, in the applications and examples it will be assumed that the fields are separable.

We regard the optical network as a linear system with a stochastic input. We are concerned with the question of what the properties of the output field are for a given input field. Strictly speaking, by "input field" we mean the actual guided field that is excited by the source, and not the source field itself. This distinction is, in principle, necessary, since the properties of the source field may be different from the properties of the excited field. For instance, even a perfectly plane-polarized wave will generally excite both polarization modes in an optical fiber, since the polarization modes of the fiber are not planarly polarized perfectly. However, the relevant statistical characteristics of the excited guided field can be taken to be the same as those of the exciting source for most practical purposes. Therefore, to simplify the nomenclature, we will disregard the distinction between the source and input fields.

In order to present the statistical description of the guided fields we will use some concepts and results from the stochastic processes theory. The theory of stochastic processes deals with experimental situations in which the outcome of the experiment is not a single or a finite set of numbers, but rather a function x(t). Quite often, the argument of the resulting function is time, and it is therefore denoted by t. If all experimental parameters were under control, then the outcome of a given experiment would be unique. The theory of stochastic processes provides a framework for the study of experiments in which the outcome of a given experiment is not unique, and is subject to the statistical fluctuations of one or more uncontrolled experimental parameters. In these circumstances, we refer to the experiment as a stochastic process, and to its particular outcome x(t) as a sample function. In general, x(t) may be a complex or even a vector function. In our case, the sample functions will be Jones vectors, which are complex vector functions.

We will review the relevant concepts of the stochastic processes theory in the first two sections of this chapter. To make this review more intelligible, we will start with real stochastic processes, and then discuss the proper generalizations for the case of complex processes. Some well-known results will be quoted without proofs, which are readily available in the literature. Instead, we will present a few examples to illustrate the various assertions. To further simplify the presentation we will ignore the distinction that is being made between being stationary in the strict or in the wide sense, and a strictly stationary process will be called simply "stationary." Analogously, we will also disregard the distinction between strict- and wide-sense cyclostationary processes for the same reason.

8.1 REAL STOCHASTIC PROCESSES

8.1.1 The Characterization of a Real Stochastic Process

For all practical purposes, a *stochastic process* is uniquely defined by an infinite set of *probability distribution functions* [1]. These distribution functions are classified by their order, which is related to the number of their arguments. The general definition of the Nth order probability distribution function is

$$F^{(N)}(x_1,x_2,\dots,x_N;\ t_1,t_2,\dots,t_N) = P\{\mathrm{x}(t_1) \le x_1,\ \mathrm{x}(t_2) \le x_2,\dots,\ \mathrm{x}(t_N) \le x_N\} \tag{8.1}$$

where $P\{condition\}$ denotes the probability that the specified condition is satisfied, and x(t) is a *sample function* of the process. We can equivalently specify the stochastic process by the corresponding *probability density functions* $f^{(N)}(x_1, x_2, \dots, x_N;\ t_1, t_2, \dots, t_N)$:

$$f^{(N)}(x_1,x_2,\dots,x_N;\ t_1,t_2,\dots,t_N) = \frac{\partial^N F^{(N)}(x_1,x_2,\dots,x_N;\ t_1,t_2,\dots,t_N)}{\partial x_1 \partial x_2 \dots \partial x_N} \tag{8.2}$$

Let us introduce the N-dimensional vectors **x** and η whose kth element is $x(t_k)$ and $\langle x(t_k)\rangle$, respectively. If $f^{(N)}(x_1, x_2, ..., x_N; t_1, t_2, ..., t_N)$ is in the form

$$f^{(N)}(x_1,x_2,...,x_N;\, t_1,t_2,...,t_N) = \frac{1}{\sqrt{(2\pi)^N \Delta}} \exp\left[-\tfrac{1}{2}(x-\eta)C^{-1}(x-\eta)^T\right] \quad (8.3)$$

for all N, where C is a positive definite matrix that depends in general on $t_1, t_2, ..., t_N$ (the covariance matrix), and Δ its determinant, then we say that the stochastic process obeys *Gaussian statistics*. Since a linear combination of Gaussian random variables is a Gaussian random variable itself, it follows that a linear combination of Gaussian stochastic processes is also a Gaussian stochastic process [1]. (By a "linear combination of stochastic processes" we mean a process whose sample functions are given linear combinations of the sample functions of the original processes.)

With the help of the probability density functions, we can compute the *ensemble average* (expectation value) of various quantities. A general quantity, the average of which we may wish to compute, is a function U that depends on N arguments, which are the values of $x(t)$ at a specified sequence of times $t_1, t_2, ..., t_N$. The average will be denoted by $\langle U[x(t_1),x(t_2),...,x(t_N)]\rangle$, and computed by the well known rule of the probability theory:

$$\langle U[x(t_1),x(t_2),...,x(t_N)]\rangle = \int d^N x\, U(x_1,x_2,...,x_N) f^{(N)}(x_1, x_2,...,x_N;\, t_1,t_2,...,t_N) \quad (8.4)$$

The result is, of course, a function of $t_1, t_2, ..., t_N$. Of particular interest are the ensemble averages of products of sample function values at different points of time:

$$G^{(N)}(t_1, t_2, ..., t_N) = \langle x(t_1)x(t_2) \ldots x(t_N)\rangle \quad (8.5)$$

These averages are called the Nth order *correlation functions* of the stochastic process. It is often more convenient to characterize the process in terms of its correlation functions; they are usually more accessible to experimental measurement and interpretation than the probability functions. To simplify the notation, we will from now on drop the superscript denoting the order of the correlation functions, since it can always be deduced from the number of arguments.

8.1.2 Stationary and Cyclostationary Processes and Their Power Spectral Densities

An important class of stochastic processes is the class of the *stationary processes*. A stochastic process will be called stationary if (and only if)

$$G(t_1, t_2, ..., t_N) = G(t_1 + \tau, t_2 + \tau, ..., t_N + \tau) \quad (8.6)$$

for any τ and N. From (8.6) we may conclude that the statistical properties of a stationary process are invariant under an arbitrary shift in the origin of the time axis. For stationary processes there exists a simple relation between its power spectrum and its second-order correlation function.

The *instantaneous average power* of a real stochastic process is defined as $\langle x^2(t)\rangle$. According to (8.5)

$$\langle x^2(t)\rangle = G(t,t) \tag{8.7}$$

which by (8.6) is independent of time for a stationary process. The second-order correlation function $G(t_1,t_2)$ of such a process will depend only on the difference $t_1 - t_2$. For stationary processes, therefore, we can introduce a single-argument second-order correlation function $R(\tau)$:

$$R(\tau) = G(t + \tau, t) \tag{8.8}$$

Two random variables x and y are statistically independent if

$$\langle xy\rangle = \langle x\rangle\langle y\rangle$$

If $x(t)$ represents the outcome of a physical stochastic process, then $x(t_1)$ and $x(t_2)$ become statistically independent as $|t_1 - t_2| \to \infty$. For such processes it is convenient to introduce a *correlation time*, τ_c, which is a qualitative measure of the "statistical persistence" time of the process. In other words, we will regard $x(t_1)$ and $x(t_2)$ as statistically independent for all practical purposes if $|t_1 - t_2|/\tau_c \gg 1$. Consequently, for physical stationary processes, we have

$$R(\tau) \to \langle x\rangle^2$$

for $|\tau|/\tau_c \to \infty$. We will not specify τ_c precisely, since this quantity will be used only for qualitative estimations. In an optical context, τ_c is called *coherence time*, a name we will use from now on.

We define the *power spectrum* $P(\nu)$ of a stationary stochastic process as the Fourier transform of $R(t)$ [1]:

$$P(\nu) = \int dt e^{2\pi i\nu t} R(t) = R(\nu) \tag{8.9}$$

From the general properties of Fourier transforms, it follows that if R(t) is characterized by a finite coherence time τ_c, then $P(\nu)$ has a characteristic width f_c, such that

$$f_c \sim \frac{1}{\tau_c}$$

We will call f_c the process *linewidth*. The relation between R(t) and $P(\nu)$ is portrayed schematically in Figure 8.1 below. The power spectrum may contain a singularity at $f = 0$ if R(∞) does not vanish.

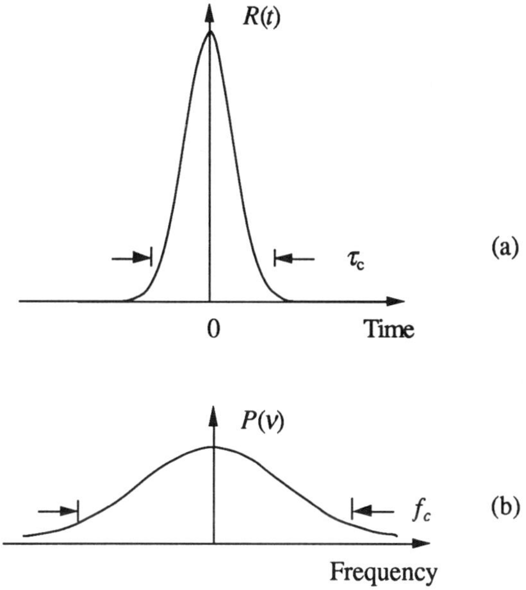

Figure 8.1 A schematic illustration of the Fourier transform relation between the correlation function and the power spectrum: (a) the correlation function; (b) the power spectrum.

It may be shown that $P(\nu) \geq 0$ and that it represents the average signal obtained by filtering the sample functions $x^2(t)$ with a linear, infintesimally narrow filter centered at ν [1]. It is also common to *define* the power spectrum of a stochastic process as the ensemble average of the Fourier transform of $x^2(t)$. In this case, (8.9) is referred to as the *Wiener-Khinchin theorem*.

We now turn to consider a more general class of stochastic processes, the class of the *cyclostationary* processes [2]. These are processes that satisfy a discrete version of (8.6), namely,

$$G(t_1, t_2, ..., t_N) = G(t_1 + mT, t_2 + mT, ..., t_N + mT) \tag{8.10}$$

for a given T and any integer m. We will refer to T as the *period* of the cyclostationary process. It is not possible to use (8.9) directly for the computation of the power spectrum of a cyclostationary process. However, for any given cyclostationary process with sample functions $x(t)$, it is possible to define a stationary process by replacing each $x(t)$ with a family of sample functions $x'(t)$ given by

$$x'(t) = x(t - \theta) \tag{8.11}$$

where θ is an independent random variable uniformly distributed in the interval $[0, T]$ [1]. The stationary process represented by $x'(t)$ is called the *shifted process*, and it possesses a power spectrum $P'(\nu)$ which can be computed from (8.9). It can be shown that the power that is obtained by filtering the sample functions $x^2(t)$ of the original (cyclostationary) process with an infinitesimally narrow linear filter centered at ν is given by $P'(\nu)$ [1]. We may therefore define the power spectrum of the original process as being equal to $P'(\nu)$. The correlation function $R'(\tau)$ of the shifted process is related to the second-order correlation function of the original cyclostationary process by

$$R'(\tau) = \frac{1}{T}\int_0^T d\theta \, G(\tau + \theta, \theta) \tag{8.12}$$

We note that a stationary process is also a cyclostationary process with an arbitrary period. The hierarchy of the various stochastic processes introduced here is shown graphically in Figure 8.2.

Throughout this book we will deal mostly with processes that are either cyclostationary or stationary and possess a finite coherence time. The generation of the optical field by CW sources is normally a stationary stochastic process. We will need the concept of cyclostationary processes to treat the output fields of time-periodic networks.

8.1.3 A Harmonic Process With a Random Amplitude (Example)

It is instructive to consider a simple example, which will justify the definition (Eq. (8.9)) of the power spectrum as well as our claims regarding the properties of cyclostationary processes. Suppose that a certain stochastic process generates the following family of sample functions $x(t)$:

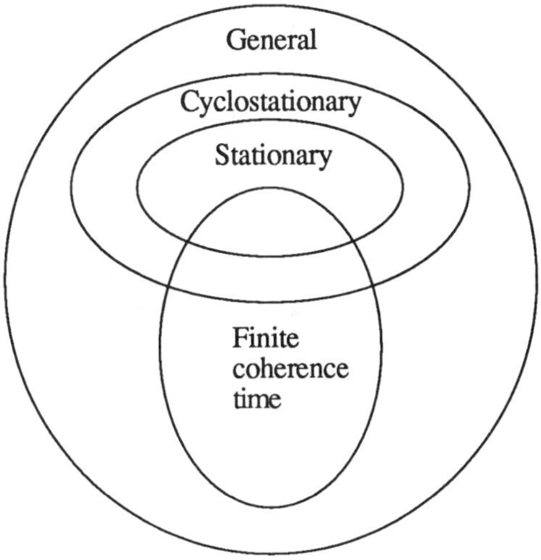

Figure 8.2 The hierarchy of the various categories of stochastic processes.

$$\mathrm{x}(t) = a \cdot \cos(2\pi\nu_0 t) \tag{8.13}$$

where a is a random variable and ν_0 a fixed frequency. This process is cyclostationary with a period $T = 1/\nu_0$. The corresponding shifted process $\mathrm{x}'(t)$ is

$$\mathrm{x}'(t) = a \cdot \cos[2\pi\nu_0(t-\theta)]$$

where θ is a random variable uncorrelated with a and uniformly distributed in the interval $[0, T]$. The correlation function $\mathrm{R}'(\tau)$ of the shifted process is given by

$$\begin{aligned}\mathrm{R}'(\tau) &= \langle \mathrm{x}'(t+\tau)\mathrm{x}'(t)\rangle \\ &= \langle a^2\rangle\langle \cos[2\pi\nu_0(t+\tau-\theta)]\cos[2\pi\nu_0(t-\theta)]\rangle \\ &= \frac{1}{2}\langle a^2\rangle\{\langle\cos[2\pi\nu_0(2t+\tau-2\theta)]\rangle + \langle\cos(2\pi\nu_0\tau)\rangle\} \\ &= \frac{1}{2}\langle a^2\rangle\cos(2\pi\nu_0\tau)\end{aligned}$$

By (8.7), the average power <P> is given by

$$\langle \mathrm{P}\rangle = R'(0) = \frac{1}{2}\langle a^2\rangle$$

and the power spectrum $P(\nu)$ by

$$P(\nu) = \frac{1}{4}\langle a^2\rangle \int d\tau e^{2\pi i\nu\tau}[\exp(2\pi i \nu_0 \tau) + \exp(-2\pi i \nu_0 \tau)]$$
$$= \frac{1}{4}\langle a^2\rangle [\delta(\nu + \nu_0) + \delta(\nu - \nu_0)]$$

Thus, the power spectrum consists of two pure lines centered at ν_0 and $-\nu_0$. The original amplitude a of the real process is divided equally between these two lines, so the total average power in each line equals $\langle a^2\rangle/4$, as was derived above. The symmetry of the power spectrum with respect to the origin of the frequency axis is characteristic of real processes. An illustration of this example appears in Figure 8.3.

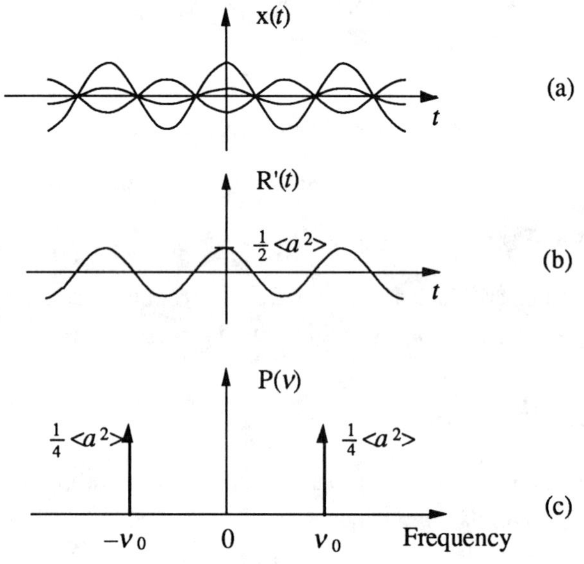

Figure 8.3 A purely harmonic stochastic process with a random amplitude: (a) the sample functions; (b) the correlation function of the shifted process; (c) the power spectrum.

8.1.4 Ensemble Averages, Time Averages, and Ergodicity

We conclude our brief review of real stochastic processes with a comment regarding ergodicity. Even though we do not need this concept to develop our theory, it is worth mentioning since, in a sense, it provides the most direct link between the stochastic processes theory description and the experimental reality. We will explain this statement by way of illustration. Let us assume that we wish to study the statistics of errors per bit in a digital communication system. Suppose that the receiver receives a signal that represents a string of, say, 10^6 bits. The received waveform will generally contain a component of additive random noise, so it can be considered as the outcome of a random process. Strictly speaking, the reception of the signal as a whole represents the outcome of a *single* experiment. Since the theory of probability can be applied only to the outcome of a large number of experiments, it appears that we cannot use the tools of the stochastic processes theory to interpret the statistics of errors in our example. This is where the ergodicity concept comes to our help. If the correlation time of the underlying process is much shorter than the duration of a bit, then instead of regarding the whole waveform as the outcome of a single experiment, we may regard the signal segments representing the various bits as the outcomes of 10^6 independent experiments. Furthermore, we assume that the "single bit" sample functions have the same statistics as the original process which generated the long 10^6 bit waveform. Adopting this point of view, we effectively replace a single experiment with a large set of independent experiments, for the study of which we can invoke statistical tools. Loosely speaking, the assumption that short, successive segments of a single, long sample function have (as an ensemble) the same statistical properties as the original ensemble of long sample functions is the ergodicity assumption. Thus, for an ergodic process the time average (which is the quantity usually accessible to experiment) equals the ensemble average (which is the quantity accessible to statistical treatment). In this book we will deal with ensemble averages only, and rely on the ergodicity assumption to relate them to the experiment.

8.2 COMPLEX STOCHASTIC PROCESSES

8.2.1 Characterization of a Complex Stochastic Process

When the sample functions are complex, we say that the process that generated them is a *complex* stochastic process. A complex sample function $\mathrm{z}(t)$ may be expressed in terms of two real sample functions $\mathrm{x}(t)$ and $\mathrm{y}(t)$:

$$\mathrm{z}(t) = \mathrm{x}(t) + i\mathrm{y}(t)$$

The statistics of a complex stochastic process is determined by the *joint* probability distribution functions

$$F^{(N)}(x_1,x_2,\ldots,y_1,y_2,\ldots;\ t_1,t_2,\ldots,s_1,s_2,\ldots)$$
$$= P\{\mathrm{x}(t_1) \le x_1,\ \mathrm{x}(t_2) \le x_2,\ \ldots,\ \mathrm{y}(s_1) \le y_1,\ \mathrm{y}(s_2) \le y_2,\ \ldots\}$$

The correlation functions that will be of interest to us in the case of complex stochastic processes are

$$G(t_1, t_2, ..., t_{2N}) = <z(t_1)z^*(t_2) \; ... \; z(t_{2N-1})z^*(t_{2N})> \tag{8.14}$$

i.e., they contain an equal number of regular and complex conjugate values of the sample function. These correlation functions are therefore always of even order. We will regard the complex process as stationary if

$$G(t_1 + \tau, t_2 + \tau, ..., t_{2N} + \tau) = G(t_1, t_2, ..., t_{2N}) \tag{8.15}$$

for any τ. Similarly, we will call such a process cyclostationary with a period T if

$$G(t_1 + mT, t_2 + mT, ..., t_{2N} + mT) = G(t_1, t_2, ..., t_{2N}) \tag{8.16}$$

for any integer m. A complex process is Gaussian if the joint statistics of the processes represented by $x(t)$ and $y(t)$ is Gaussian.

The instantaneous average power $<P>$ of a complex process is defined by

$$<P> = <|z(t)|^2> = G(t,t) \tag{8.17}$$

By (8.15), for a stationary process, $<P>$ is time-independent and is equal to G(0,0). By complete analogy to the real case, we define for a stationary complex process a single-argument correlation function $R(\tau)$:

$$R(\tau) = G(t + \tau, t) \tag{8.18}$$

and the power spectrum $P(\nu)$ is defined by (8.9):

$$P(\nu) = \int dt e^{2\pi i \nu t} R(t) = R(\nu)$$

8.2.2 A Superposition of Pure Spectral Lines With Random Amplitudes (Example)

Let us reexamine the simple example we considered in Section 8.1 with a slight generalization. Suppose that we are given a complex stochastic process that generates the following family of sample functions:

$$z(t) = \sum_j a_j \exp(-2\pi i \nu_j t) \tag{8.19}$$

where the coefficients a_j are statistically independent, zero-mean complex random variables. Equation (8.19) describes a random superposition of pure spectral lines, with the jth line having an average power equal to $\langle |a_j|^2 \rangle$. The correlation function $R(\tau)$ of this process is found to be

$$R(\tau) = \sum_j \langle |a_j|^2 \rangle \exp(-2\pi i \nu_j \tau)$$

Applying (8.9), we get

$$P(\nu) = \sum_j \langle |a_j|^2 \rangle \delta(\nu - \nu_j) \tag{8.20}$$

If we now integrate $P(\nu)$ over a given frequency interval $[\nu_1, \nu_2]$, we obtain

$$\int_{\nu_1}^{\nu_2} d\nu P(\nu) = \sum \langle |a_j|^2 \rangle$$

where the summation is performed over all frequencies ν_j within the interval $[\nu_1, \nu_2]$. Thus, the integral of $P(\nu)$ over a given frequency interval is the sum of the powers of the spectral lines, the frequencies of which are contained in that interval, in accord with what could be expected.

8.3 THE SECOND-ORDER FIELD CORRELATION FUNCTIONS AND THE OPTICAL FIELD POWER SPECTRUM

8.3.1 Statistical Description of Jones Vectors

We intend to apply the stochastic processes theory to the study of the statistical properties of the guided fields. As we have seen, a guided field is described by a two-component complex vector, which we have called a Jones vector. We regard each one of its two components as a sample function of a stochastic process. Here we will be interested in the analysis of narrowband fields. A Jones vector $\mathbf{A}(t)$ will be considered narrowband if it can be cast in the form

$$\mathbf{A}(t) = \mathbf{a}(t)e^{-2\pi i \nu t} \tag{8.21}$$

where $\mathbf{a}(t)$ is a complex vector sample function of a random process with a power spectrum range much smaller than ν. The frequency ν is in the optical case on the order of 100 THz, so that even if the spectrum of $\mathbf{a}(t)$ spans a spectral range as large as 100 GHz, the field can be still considered narrowband to a very good approximation. We will refer to $\mathbf{a}(t)$ as the *complex amplitude* of the Jones vector $\mathbf{A}(t)$.

In this book we will be concerned exclusively with optical sources in which the complex amplitude is stationary. (By saying that a vector process is stationary or cyclostationary, we mean that its components are mutually stationary or cyclostationary, respectively.) It is easy to show that if the complex amplitude is stationary, then (8.21) is stationary as well (in the sense of (8.15)). However, the representation in (8.21) is not physical, since no known physical instrument can measure or control the absolute value of an optical field phase. Actually, the complete loss of phase information is inherent in a quantum intensity detection process. To account for that fact, we represent optical fields by the sample functions

$$\mathbf{A}'(t) = \mathbf{a}(t)e^{-2\pi i \nu (t-\theta)} \tag{8.22}$$

where θ is a random variable statistically independent from $\mathbf{a}(t)$ and uniformly distributed in the interval $[0, 1/\nu]$. The process represented by (8.22) may be regarded as the combination of the field generation and measurement processes. From now on we will assume that all Jones vectors are sample functions of this combined process, and include automatically the averaging over θ in the ensemble averages. Since $\langle e^{2\pi i \nu \theta}\rangle = 0$, it follows that

$$\langle \mathbf{A}(t)\rangle = \langle \mathbf{a}(t)\rangle\langle e^{-2\pi i \nu (t-\theta)}\rangle = 0 \tag{8.23}$$

or, in other words, the average of a narrowband Jones vector vanishes. Also, if the components $A_1(t)$ and $A_2(t)$ of the Jones vector are uncorrelated, then their cross correlation vanishes:

$$\langle A_1(t_1)A_2(t_2)\rangle = \langle A_1(t_1)\rangle\langle A_2(t_2)\rangle = 0 \tag{8.24}$$

8.3.2 The Second-Order Statistics of Guided Fields

The second-order statistics of the guided field is described by the so called *coherency matrix* $\mathbf{G}(t_1,t_2)$ [3]:

$$G_{jk}(t_1,t_2) = <A_j(t_1)A_k^*(t_2)> \tag{8.25}$$

We note that it is possible to define the coherency matrix using vector notation, namely,

$$\mathbf{G}(t_1,t_2) = <\mathbf{A}(t_1)\mathbf{A}^\dagger(t_2)> \tag{8.26}$$

From this definition it can be seen that

$$\mathbf{G}^\dagger(t_1,t_2) = <\mathbf{A}(t_2)\mathbf{A}^\dagger(t_1)> = \mathbf{G}(t_2,t_1) \tag{8.27}$$

so that $\mathbf{G}(t,t)$ is hermitian. It is important to note that in a degenerate waveguide, like a perfectly symmetric fiber, there is a certain degree of arbitrariness in the definition of the coherency matrix. Indeed, in a degenerate waveguide, any two orthogonal combinations of the planar polarization modes (like the circularly polarized modes) can be taken as a basis for the Jones vector definition. Thus, if $\boldsymbol{U}$ is a unitary matrix, then

$$\mathbf{G}'(t_1, t_2) = <\boldsymbol{U}\mathbf{A}(t_1)\mathbf{A}^\dagger(t_2)\boldsymbol{U}^\dagger> = \boldsymbol{U}\mathbf{G}(t_1,t_2)\boldsymbol{U}^\dagger$$

can be regarded also as the coherency matrix of the same field.

We define the frequency-domain representation $\boldsymbol{G}(\nu_1,\nu_2)$ of the coherency matrix as

$$\begin{aligned} \boldsymbol{G}(\nu_1,\nu_2) &= \int d^2t \exp[2\pi i(\nu_1 t_1 - \nu_2 t_2)]\mathbf{G}(t_1,t_2) \\ &= <\boldsymbol{A}(\nu_1)\boldsymbol{A}^\dagger(\nu_2)> \end{aligned} \tag{8.28}$$

We call $\boldsymbol{G}(\nu_1,\nu_2)$ the frequency-domain representation rather than the Fourier transform of $\mathbf{G}(t_1,t_2)$ since in the standard Fourier transform definition, all terms in the exponent appear with a positive sign. The choice of the negative sign in the second term on the exponent allows us to express $\boldsymbol{G}(\nu_1,\nu_2)$ in terms of the product $\boldsymbol{A}(\nu_1)\boldsymbol{A}^\dagger(\nu_2)$, as shown in the bottom line of (8.28).

For a stationary process, $\mathbf{G}(t_1,t_2)$ depends only on the difference $t_1 - t_2$. In this case, we can reduce the double integral in (8.28) to a single integral by introducing new variables

$$s_1 = t_1 - t_2$$

$$s_2 = t_2$$

In terms of these variables, (8.28) is expressed as

$$\begin{aligned} G(\nu_1, \nu_2) &= \int d^2s \exp\{2\pi i[\nu_1(s_1 + s_2) - \nu_2 s_2]\}\mathbf{R}(s_1) \\ &= R(\nu_1)\delta(\nu_1 - \nu_2) \end{aligned} \tag{8.29}$$

where we have introduced the single-argument coherency matrix

$$\mathbf{R}(s_1) = \mathbf{G}(t_1, t_2) \tag{8.30}$$

Let us note that

$$\mathbf{R}^{\dagger}(s) = \mathbf{R}(-s) \tag{8.31}$$

8.3.3 Optical Field Intensity and Spectrum

We turn now to a discussion of the optical field power and the optical field spectrum. It is customary to refer to the optical field power as the optical field intensity, and to denote it with I(t). Its average is derived from (2.3):

$$\langle \mathrm{I}(t) \rangle = \sum_j \langle |\mathrm{A}_j(t)|^2 \rangle \zeta \tag{8.32}$$

From now on we will drop the irrelevant power unit ζ, assuming that its numerical value is 1. From (8.25) and (8.32) it follows that

$$\langle \mathrm{I}(t) \rangle = \mathrm{Tr}[G(t,t)] \tag{8.33}$$

This definition of the average intensity is independent of the particular polarization mode basis, since for any unitary matrix U

$$\mathrm{Tr}[U\mathbf{G}(t, t)U^{\dagger}] = \mathrm{Tr}[U^{\dagger}U\mathbf{G}(t, t)] = \mathrm{Tr}[\mathbf{G}(t, t)]$$

For a stationary field the average field intensity is time-independent, and may be expressed in terms of the single parameter coherency matrix (8.30):

$$<I> = \mathrm{Tr}[\mathbf{R}(0)] \tag{8.34}$$

In view of (8.32), the proper generalization of (8.9) in the optical field case is

$$P(\nu) = \int dt e^{2\pi i \nu t} \mathrm{Tr}[\mathbf{R}(t)] = \mathrm{Tr}[\boldsymbol{R}(\nu)] \tag{8.35}$$

The optical field power spectrum $P(\nu)$ can be measured with a tunable, narrowband optical filter (a spectrometer). From the well-known properties of Fourier transforms, it is evident that $P(\nu)$ is significantly nonzero in a frequency interval of length f_c, centered at the optical frequency, and it decays to zero outside this interval, as shown schematically in Figure 8.1. We will refer to $P(\nu)$ also as the *field lineshape*. Finally, we note that the average intensity $<I>$ can also be expressed as

$$<I> = \int d\nu \mathrm{Tr}[\boldsymbol{R}(\nu)] \tag{8.36}$$

8.3.4 The Separable Field

The matrix $\boldsymbol{R}(\nu)$ is hermitian, since

$$\boldsymbol{R}^{\dagger}(\nu) = \int dt e^{-2\pi i \nu t} \mathbf{R}^{\dagger}(t) = \int dt e^{2\pi i \nu t} \mathbf{R}(t) = \boldsymbol{R}(\nu) \tag{8.37}$$

where we have used (8.31). Therefore, it is always possible to find a polarization mode basis in which $\boldsymbol{R}(\nu)$ will appear diagonal for a *given* optical frequency ν_0. In general, $\boldsymbol{R}(\nu)$ will not be diagonal in this basis for optical frequencies other than ν_0. This fact makes the analysis of general optical fields considerably more complicated than the corresponding analysis of scalar signals (like a microwave field in a strictly single-mode waveguide). In practice, however, it is normally reasonable to assume that the field is a superposition of two orthogonal and statistically independent polarization modes (which do not have to be planar). We will call such a field a *separable field*, and the source that generates such a field a *separable source*. Since the cross correlation of two independent Jones vector components vanishes, choosing the two independent polarization modes as the polarization modes basis, the correlation matrix $\boldsymbol{R}(\nu)$ of a separable field takes the form

$$\boldsymbol{R}(v) = \begin{pmatrix} P_1(v) & 0 \\ 0 & P_2(v) \end{pmatrix} \tag{8.38}$$

where $P_1(v)$ and $P_2(v)$ are the power spectra of the field in the corresponding polarization modes.

Both the perfectly polarized and unpolarized fields are special cases of the separable field. In the first case, if one of the basis polarization modes is chosen to coincide with the mode in which the source is polarized, then the correlation matrix is in the form

$$\boldsymbol{R}(v) = \begin{pmatrix} P_1(v) & 0 \\ 0 & 0 \end{pmatrix} \tag{8.39}$$

In the perfectly unpolarized field case, for an *arbitrary* choice of the basis polarization modes, the correlation matrix has the form

$$\boldsymbol{R}(v) = \frac{1}{2}P(v)\boldsymbol{I} \tag{8.40}$$

In both cases, there exists a polarization mode basis in which $\boldsymbol{R}(v)$ is diagonal, which is our definition of a separable field. The analysis of a separable field problem is often reduced to two independent analyses of polarized field problems.

The hierarchy of the various polarization models of guided optical fields is shown graphically in Figure 8.4.

8.4 REPRESENTATION OF A REAL JONES VECTOR BY A COMPLEX ONE

8.4.1 The Real Jones Vector and Its Associated Complex Jones Vector

Complex Jones vectors are much more convenient for analytical treatment than real Jones vectors. However, complex Jones vectors do not exist in reality since optical fields are real. Fortunately, in the narrowband field case there is a well-known and simple way to associate a complex Jones vector with a real one, in such a way that the properties of the real field can be easily derived from the properties of the corresponding complex field. The common approach is therefore to perform all the analysis with the complex Jones vectors, and to use certain correspondence rules to derive the relevant results for the real field. In this subsection we discuss the correspondence rules that are relevant to the second-order field statistics.

The frequency-domain representation of real processes is always symmetric with respect to the origin. For this reason, in order to describe a real stochastic process, the positive frequencies are sufficient. However, the frequency-domain representation of complex processes is not symmetric in general, and therefore both positive and negative frequencies are necessary for the description of such processes. In some contexts, a representation that uses both positive and negative frequencies is called *two-sided representation.*

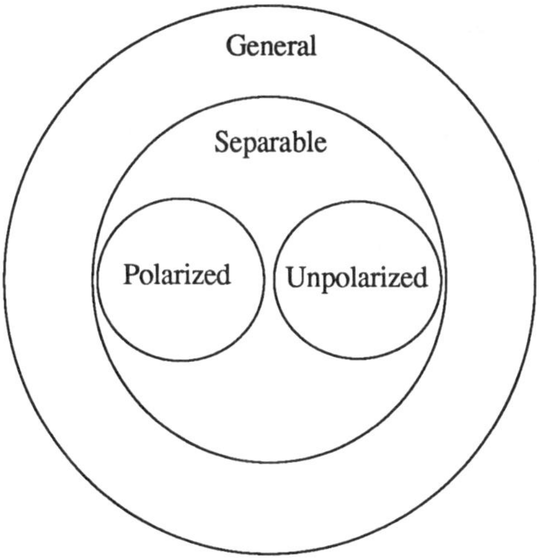

Figure 8.4 The hierarchy of the various polarization models of guided optical fields.

Analogously to (8.21), a real Jones vector $\mathbf{A}(t)$ corresponding to a narrowband real field can be represented as

$$\mathbf{A}(t) = \mathbf{a}_1(t)\cos(2\pi\nu t) + \mathbf{a}_2(t)\sin(2\pi\nu t) \tag{8.41}$$

where the range of the power spectra of $\mathbf{a}_1(t)$ and $\mathbf{a}_2(t)$ is much smaller than the optical frequency ν. Analogously to (8.22), we represent the uncertainty in the phase measurement by introducing an independent random variable θ uniformly distributed in the interval $[0, 1/\nu]$:

$$\mathbf{A}(t) = \mathbf{a}_1(t)\cos[2\pi\nu(t-\theta)] + \mathbf{a}_2(t)\sin[2\pi\nu(t-\theta)] \tag{8.42}$$

We associate with $\mathbf{A}(t)$ the complex Jones vector $\mathbf{B}(t)$:

$$\mathbf{B}(t) = \mathbf{b}(t)e^{-2\pi i\nu(t-\theta)} \tag{8.43}$$

where

$$\mathbf{b}(t) = \frac{1}{\sqrt{2}}[\mathbf{a}_1(t) - i\mathbf{a}_2(t)] \tag{8.44}$$

It is easily verified that

$$\mathbf{A}(t) = \frac{1}{\sqrt{2}}[\mathbf{B}(t) + \mathbf{B}^*(t)] \tag{8.45}$$

The numerical factor $1/\sqrt{2}$ has been chosen in order to make the *average* intensities of the real process and its associated complex process equal. Indeed,

$$<\mathrm{I}_A> = <\mathbf{A}^2> = \frac{1}{2}(<\mathbf{a}_1^2> + <\mathbf{a}_2^2>)$$

and

$$<\mathrm{I}_B> = <|\mathbf{B}|^2> = \frac{1}{2}(<\mathbf{a}_1^2> + <\mathbf{a}_2^2>) = <\mathrm{I}_A>$$

8.4.2 The Relation Between Coherency Matrices and Power Spectra of Real and Complex Processes

Let us compute now the coherency matrix $\mathbf{G}_A(t_1,t_2)$ of $\mathbf{A}(t)$:

$$\mathbf{G}_A(t_1,t_2) = <\mathbf{A}(t_1)\mathbf{A}^{\mathrm{T}}(t_2)>$$

$$= \frac{1}{2}<[\mathbf{B}(t_1) + \mathbf{B}^*(t_1)][\mathbf{B}^{\mathrm{T}}(t_2) + \mathbf{B}^{\dagger}(t_2)]>$$

Noting that the average of all terms containing θ vanishes, we are left with

$$\mathbf{G}_A(t_1,t_2) = \frac{1}{2}[\mathbf{G}_B(t_1,t_2) + \mathbf{G}_B^*(t_1,t_2)] \tag{8.46}$$

where $\mathbf{G}_B(t_1,t_2)$ is the coherency matrix of $\mathbf{B}(t)$:

$$\mathbf{G}_B(t_1,t_2) = <\mathbf{B}(t_1)\mathbf{B}^\dagger(t_2)> \tag{8.47}$$

Equation (8.46) represents the correspondence rule between the coherency matrices of a real process and its associated complex process. We turn now to derive the rule for the correspondence between the power spectra of these processes. From (8.46) it follows that

$$\boldsymbol{R}_A(\nu) = \frac{1}{2}\,[\boldsymbol{R}_B(\nu) + \boldsymbol{R}_B{}^*(-\nu)]$$

Since $\boldsymbol{R}_B(\nu)$ is hermitian, we obtain

$$P_A(\nu) = \mathrm{Tr}[\boldsymbol{R}_A(\nu)] = \frac{1}{2}\,\mathrm{Tr}[\boldsymbol{R}_B(\nu) + \boldsymbol{R}_B(-\nu)] = \frac{1}{2}\,[P_B(\nu) + P_B(-\nu)] \tag{8.48}$$

This equation indicates that the power of the real process is split equally between two symmetric spectral components. The total average power in each component is half the average power of the associated complex process, but the total average powers of both processes are the same. The relation between $P_A(\nu)$ and $P_B(\nu)$ is illustrated schematically in Figure 8.5.

From (8.46) and (8.48) it is seen that the coherency matrix and the power spectrum of the real process are very simply related to the corresponding quantities of the associated complex process. This justifies the use of the complex representation for the analytic treatment, at least as far as second-order statistical properties are concerned. In the next chapter we will see that the complex representation is useful for the computation of the relevant fourth-order statistical properties as well.

8.5 STATISTICAL MODEL OF AN AMPLITUDE-STABILIZED LASER

In many occasions the amplitude of a laser source is stabilized to reduce the noise. A convenient and useful model of an amplitude-stabilized laser field is the *random-phase* model [4]. In this model, we assume that the instantaneous optical frequency ν_j of the jth polarization mode is time-dependent, and cast it in the form

$$\nu_j(t) = \nu_0 + \Delta\nu_j(t) \tag{8.49}$$

where ν_0 is the (time-independent) average frequency, and the frequency deviation $\Delta\nu_j(t)$ is a sample function of a stochastic process with a zero mean. We will assume that this process is Gaussian, an assumption that can be justified in practice by physical arguments.

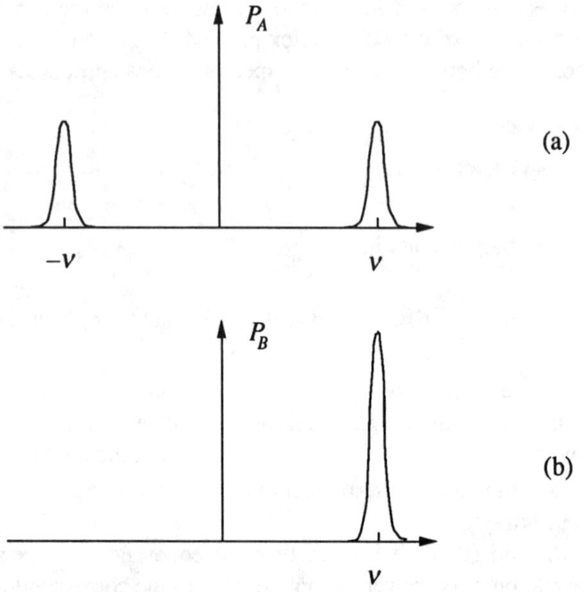

Figure 8.5 The power spectrum of a real process and its associated complex process (schematic): (a) the power spectrum of the real process; (b) the power spectrum of the associated complex process.

We assume that the Jones vector components $A_j(t)$ of a random-phase field are given by

$$A_j(t) = a_j \exp[-2\pi i \nu_0 (t - \theta) + i\phi_j(t)] \tag{8.50}$$

where a_1 and a_2 are complex constants. The phase $\phi_j(t)$ is expressed as a time integral of the instantaneous frequency deviation $\Delta\nu_j(t)$:

$$\phi_j(t) = 2\pi \int_0^t dt' \Delta\nu_j(t') + \phi_{j0} \tag{8.51}$$

where ϕ_{j0} is an independent random variable distributed uniformly in the interval $[0, 2\pi]$. Under these assumptions, the cross correlation between the two polarization modes vanishes, since

$$<A_1(t_1)A_2^*(t_2)> \propto <\exp(i\phi_{10})><\exp(-i\phi_{20})> = 0$$

so that the random-phase field is separable, and the analysis is reduced to the analysis of two independent polarized fields. We present here the treatment of one of the polarization components and, for simplicity, drop the subscript denoting the polarization mode.

Let us denote the correlation function of $\Delta\nu(t)$ with $G_\nu(\tau)$:

$$G_\nu(\tau) = <\Delta\nu(t+\tau)\Delta\nu(t)> \tag{8.52}$$

Note that $G_\nu(\tau)$ is an even function of τ; i.e.,

$$G_\nu(-\tau) = G_\nu(\tau) \tag{8.53}$$

The correlation function $R(\tau)$ of the random-phase field is given by

$$R(\tau) = \exp(-2\pi i\nu_0\tau)|a|^2<\exp\{i[\phi(t+\tau) - \phi(t)]\}> \tag{8.54}$$

We note that

$$\phi(t+\tau) - \phi(t) = 2\pi\int_t^{t+\tau} dt'\Delta\nu(t') \tag{8.55}$$

and therefore the difference $\phi(t+\tau) - \phi(t)$ is a Gaussian variable. In order to compute the average appearing in the right-hand side of (8.54), we can use the relation [1]

$$<\exp(ix)> = \exp[i<x> - \tfrac{1}{2}(<x^2> - <x>^2)] \tag{8.56}$$

which is true for any Gaussian random variable x. In our case,

$$<[\phi(t+\tau) - \phi(t)]> = 2\pi\int_t^{t+\tau} dt'<\Delta\nu(t')> = 0 \tag{8.57}$$

Therefore,

$$<\exp\{i[\phi(t+\tau) - \phi(t)]\}> = \exp\{-\tfrac{1}{2}<[\phi(t+\tau) - \phi(t)]^2>\} \tag{8.58}$$

From (8.51) and (8.55) we get

$$<[\phi(t+\tau)-\phi(t)]^2> = 4\pi^2 \int_t^{t+\tau} d^2t G_\nu(t_1-t_2)$$

$$= 4\pi^2 \int_0^{\tau} d^2t G_\nu(t_1-t_2) \tag{8.59}$$

Let us denote the last integral with $\Lambda(\tau)$:

$$\Lambda(\tau) = 4\pi^2 \int_0^{\tau} d^2t G_\nu(t_1-t_2) \tag{8.60}$$

The function $\Lambda(\tau)$ is also called the *phase structure function*. We note that since $G_\nu(t)$ is an even function, it follows that $\Lambda(\tau)$ is an even function as well. Therefore, it is sufficient to evaluate $\Lambda(\tau)$ for $\tau \geq 0$. In order to carry out this evaluation, it is useful to introduce the function rect(x) defined by

$$\text{rect}(x) = \begin{cases} 1 & \text{for } 0 \leq x \leq 1, \\ 0 & \text{otherwise.} \end{cases} \tag{8.61}$$

With the help of this function, we can rewrite (8.60) in the form

$$\Lambda(\tau) = 4\pi^2 \int d^2t \,\text{rect}(\frac{t_1}{\tau})\text{rect}(\frac{t_2}{\tau})G_\nu(t_1-t_2)$$

We now introduce the variables

$$u_1 = t_1 - t_2, \quad u_2 = t_1 + t_2$$

Using these variables, we express $\Lambda(\tau)$ in the form

$$\Lambda(\tau) = 2\pi^2 \int d^2u \,\text{rect}(\frac{u_2+u_1}{2\tau})\text{rect}(\frac{u_2-u_1}{2\tau})G_\nu(u_1) \tag{8.62}$$

The integration over u_2 may be performed analytically. Let us write

$$\mathrm{I}(u_1) = \int du_2 \, \mathrm{rect}\left(\frac{u_2 + u_1}{2\tau}\right)\mathrm{rect}\left(\frac{u_2 - u_1}{2\tau}\right) \tag{8.63}$$

To evaluate this integral, we introduce

$$y = \frac{u_2 + u_1}{2\tau} \tag{8.64}$$

so that

$$I(u_1) = 2\tau\int dy \, \mathrm{rect}(y)\mathrm{rect}\left(y - \frac{u_1}{\tau}\right) \tag{8.65}$$

Inspection of Figure 8.6 reveals that

$$\mathrm{I}(u_1) = \begin{cases} 2\tau\left(1 - \dfrac{u_1}{\tau}\right) & \text{for } |u_1| \le \tau, \\ 0 & \text{otherwise.} \end{cases} \tag{8.66}$$

(We remind the reader that we assume that $\tau \ge 0$.) Substituting this expression for $\mathrm{I}(u_1)$ in (8.64), we finally obtain

$$\Lambda(\tau) = 4\pi^2\tau\int_{-\tau}^{\tau} du\left(1 - \frac{u}{\tau}\right)\mathrm{G}_v(u) \tag{8.67}$$

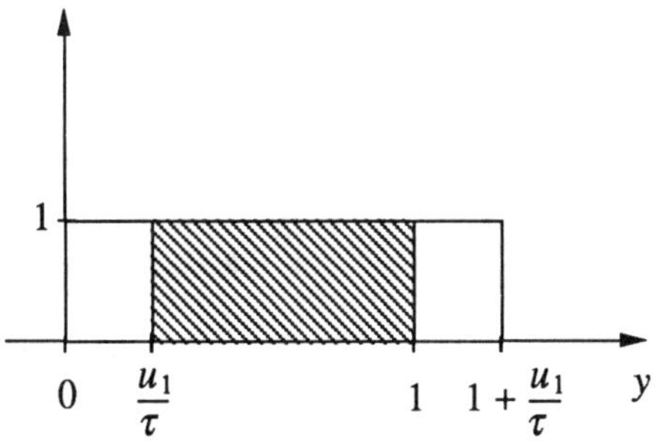

Figure 8.6 The two factors in the integrand of (8.65) and their overlap.

It is worthwhile to derive the asymptotic behavior of $\Lambda(\tau)$. For $\tau \ll \tau_c$ we may approximate $\mathrm{G}_v(u)$ in the region of the integration by $\mathrm{G}_v(0)$, to obtain

$$\Lambda(\tau) \approx 4\pi^2\tau^2 G_\nu(0) \tag{8.68}$$

In the case that $\tau \gg \tau_c$, we neglect the fraction u/τ in the first factor in the integrand, and obtain

$$\Lambda(\tau) \approx 4\pi^2|\tau| \int du G_\nu(u) \tag{8.69}$$

Note that we have to use the absolute value of τ on the right-hand side, since $\Lambda(\tau)$ must be an even function of τ.

To sum up, the coherency matrix of the random-phase field has the following form:

$$\mathbf{R}(\tau) = \exp(-2\pi i \nu_0 \tau) \begin{pmatrix} I_1 \exp\left[-\frac{1}{2}\Lambda_1(\tau)\right] & 0 \\ 0 & I_2 \exp\left[-\frac{1}{2}\Lambda_2(\tau)\right] \end{pmatrix} \tag{8.70}$$

where

$$I_j = |a_j|^2$$

is the field intensity in mode j. For model computations with the random-phase field, it is often assumed that

$$\Lambda_j(\tau) = \gamma_j|\tau| \tag{8.71}$$

with γ_j being a real constant, related to the field linewidth f_c. This form for the phase structure function describes a random-walk-type diffusion [4]. It should be noted that (8.71) contradicts (8.68), according to which $\Lambda(\tau)$ should behave as τ^2 for $\tau \to 0$. This contradiction may result in certain artifacts, and therefore (8.71) must be applied cautiously.

Let us compute the power spectrum for a random-phase field satisfying (8.71). From (8.35) we obtain

$$\begin{aligned} P(\nu) &= \sum_j I_j \int d\tau \exp[2\pi i \tau(\nu - \nu_0) - \tfrac{1}{2}\gamma_j|\tau|] \\ &= \sum_j \frac{4}{\gamma_j} \frac{I_j}{1 + [4\pi(\nu - \nu_0)/\gamma_j]^2} \end{aligned} \tag{8.72}$$

Consequently, the half-height linewidth of the field at each polarization mode can be seen to be $\gamma_i/2\pi$. The relatively slow decay of $P(\nu)$ as $\nu \to \infty$ (as the inverse of the second power of ν) is caused by the artificial discontinuity in the first derivative of $\Lambda(\tau)$ at $\tau = 0$.

8.6 SUMMARY

We regard the optical field as a narrowband stochastic process. Since this field is described by a two-component complex Jones vector, it is actually composed of two complex stochastic processes.

A great deal of our analysis will be based on several fundamental results and concepts from the random process theory. For the convenience of the reader, we have included a concise review of the most relevant issues at the beginning of this chapter. The concepts that were reviewed are the correlation functions, probability distribution and density functions, ensemble averages, Gaussian statistics, stationary and cyclostationary processes, and spectral power density. We have also shown the relation between the second-order correlation function and the spectral power density, which is usually referred to as the Wiener-Khinchin theorem. The optical power spectra are computed using this relation.

The actual values of the Jones vector components depend on the representation. If there exists a representation in which these components are statistically independent, we say that the field is separable. Both the polarized and the unpolarized fields are special cases of separable fields. Also, as far as we know, the optical fields that are generated by laser diodes are separable as well. The analysis of separable fields is much simpler than the analysis of a general field, and it is often reduced to two independent analyses of polarized fields.

Another type of simplification is introduced by considering complex quantities instead of real ones. We have shown that there is a simple way to associate a complex Jones vector to any given real and narrowband Jones vector. There is also a simple correspondence between the power spectra of the real and its associated complex processes. In the next chapter we will see that the complex representation is also useful for problems involving fourth-order statistics.

A narrowband Jones vector reperesents in general a nonstationary process. This is due to the periodic time dependence of the optical fields. Rigorously speaking, a narrowband Jones vector usually represents a cyclostationary process. This fact, however, is not manifested in practice, since no physical instrument can presently measure or control the *absolute* phase of an optical field. To account for that, we add a random phase to the exponential factor describing the periodic time dependence of the optical field. This operation generates a new family of sampling functions, which will usually represent a stationary field. The new family of sampling functions may be regarded as representing a combination of two processes: the generation and the measurement of an optical field.

A popular model for the description of highly coherent and amplitude-stabilized fields is the random-phase field. In this model it is assumed that the only random element in the Jones vector is the phase. By further assuming that the phase exhibits a random-walk-type diffusion process, we can obtain an explicit expression for the second-order correlation functions of the field and its lineshape.

References

[1] Papoulis, A., *Probability, Random Variables and Stochastic Processes*, New York: McGraw-Hill, 1984.

[2] Gardner N.A. and L.E. Franks, "Characterization of Cyclostationary Random Signal Processes," *IEEE Transactions on Information Theory*, Vol. IT-21, No. 1, January 1975, pp. 4-14.

[3] Wolf E., "Optics in Terms of Observable Quantities," *Nuovo Cimento*, Vol. 12, 1954, pp. 884-888.

[4] Armstrong, J.A., "Theory of Interferometric Analysis of Laser Phase Noise," *J. Opt. Soc. Am.*, Vol. 56, No. 8, August 1966, pp. 1024-1031.

Chapter 9
The Fourth-Order Statistics and the Optical Intensity Power Spectrum

In most experimental circumstances, the direct output of optical detectors is an electrical current. In an ideal, infinite-bandwidth, and noiseless detector, this current is exactly proportional to the instantaneous incident optical intensity $I(t)$. Therefore, random fluctuations in the optical intensity will induce a noise in the current of such a detector. We will refer to this noise as the *field-induced noise*. In real detectors there are, in general, other contributions to the noise, most notably the shot noise, which is inherent in any quantum detection system. The relative powers of the various noise sources depend on the experimental circumstances. In many practical cases the magnitude of the field-induced noise is significant, and sometimes even dominant. Therefore, the contribution of the field-induced noise should not be disregarded in the system noise evaluation.

Our main concern in this chapter is the investigation of the field-induced noise power spectrum. Since the field power spectrum is determined by the second-order field statistics, the field *intensity* power spectrum will be determined by the fourth-order field statistics. Therefore, to unveil the characteristics of the field-induced noise, we have to investigate the properties of the fourth-order field correlation functions.

The treatments of the field-induced noise that have appeared so far in the literature were carried out mostly in the time domain. This preference for the time-domain representation is probably due to the traditional approach to modelling of interferometric systems in optics . In general, if the goal is just to assess the noise that is induced by a given field, both time- and frequency-domain approaches may be equally suitable. However, if one is concerned with the effect of a time-independent network on the field-induced noise, there is a definite advantage in using the frequency-domain approach, since, as we have seen, the transformation of frequency-domain quantities by time-independent networks involves simple products, while the same transformation applied to time-domain quantities requires the evaluation of convolution integrals. For this reason, we invest a considerable effort in deriving the frequency-domain representation of the field intensity power spectrum, which will turn out to be very useful in the next chapter.

9.1 THE FOURTH-ORDER FIELD CORRELATION FUNCTIONS

9.1.1 General Properties

The fourth-order field correlation functions of a guided field are defined as

$$G_{nmkj}(t_1,t_2,t_3,t_4) = \langle A_n(t_1)A_m{}^*(t_2)A_k(t_3)A_j{}^*(t_4)\rangle \tag{9.1}$$

where the quantities that appear in the right-hand side are the components of the field Jones vector **A**(t). The fourth-order correlation functions are denoted by four indices, each of which can take one of two values, 1 or 2. There are therefore sixteen fourth-order correlation functions that depend on four time variables. If the field is Gaussian, then its fourth-order field correlation functions may be expressed in terms of the second-order correlation functions [1]:

$$G_{nmkj}(t_1,t_2,t_3,t_4) = G_{nm}(t_1,t_2)G_{kj}(t_3,t_4) + G_{nj}(t_1,t_4)G_{km}(t_3,t_2) \tag{9.2}$$

We see that in the Gaussian field case, all its fourth-order correlation functions are expressed in terms of the elements of its coherency matrix, of which actually only three are independent. This fact makes the Gaussian field much easier for both experimental characterization and analytic treatment.

A Gaussian field may have an arbitrary coherency matrix, and therefore also an arbitrary lineshape. Furthermore, all statistical properties of a (zero mean) Gaussian field are determined by its coherency matrix [1]. Therefore, it is possible to associate with any given field a unique Gaussian field whose coherency matrix coincides with the coherency matrix of the given field. We will refer to this field as the *equivalent Gaussian field*, and denote its fourth-order correlation functions with $\Gamma_{nmkj}(t_1,t_2,t_3,t_4)$:

$$\Gamma_{nmkj}(t_1,t_2,t_3,t_4) = G_{nm}(t_1,t_2)G_{kj}(t_3,t_4) + G_{nj}(t_1,t_4)G_{km}(t_3,t_2) \tag{9.3}$$

where the coherency matrix elements on the right-hand side are the coherency matrix elements of the given field.

The frequency-domain representation of the coherency matrix of a stationary process contains a δ function singularity, as is shown explicitly in (8.29). It is to be expected that the frequency-domain representation of the fourth-order correlation functions will contain singularities as well. These singularities must be exposed in order to derive a useful frequency-domain representation for these functions.

The key to the frequency-domain singularities is the asymptotic behavior in the time domain. We assume that the process is stationary and narrowband. Such a process possesses a finite coherence time τ_c. Our aim is to find out what are the asymptotic values of $G_{nmkj}(t_1,t_2,t_3,t_4)$ when the absolute value of the difference between any two of its arguments becomes much larger than τ_c [2].

Suppose that there exists at least one pair of time arguments, say t_p and t_q, such that $|t_p - t_q| \gg \tau_c$. Then, the Jones vector component (or its complex conjugate) having t_p as its argument becomes uncorrelated with the corresponding component (or its complex conjugate) having t_q as its argument. It may, however, remain correlated with any one of the two remaining factors comprising $G_{nmkj}(t_1,t_2,t_3,t_4)$, or even with both. However, terms containing the average of a single Jones vector component vanish (Eq. (8.23)). Therefore, if $|t_p - t_q| \gg \tau_c$ then the fourth-order correlation function must take the form

$$G_{nmkj}(t_1,t_2,t_3,t_4) = \langle A_n(t_1)A_k(t_3)\rangle\langle A_j^*(t_4)A_m^*(t_2)\rangle + \Gamma_{nmkj}(t_1,t_2,t_3,t_4) \tag{9.4}$$

where $\Gamma_{nmkj}(t_1,t_2,t_3,t_4)$ is the fourth-order correlation function of the equivalent Gaussian field. The first term contains a factor in the form

$$\langle A_n(t_1)A_k(t_3)\rangle = \langle a_n(t_1)a_k(t_3)\rangle\langle \exp[2\pi i\nu(t_1 + t_3 - 2\theta)]\rangle$$

where $\mathbf{a}(t)$ is the Jones vector complex amplitude. This factor vanishes, since it is proportional to the average of $e^{4\pi i\nu\theta}$. We may conclude, therefore, that as $|t_p - t_q|/\tau_c \to \infty$, then

$$G_{nmkj}(t_1,t_2,t_3,t_4) \to \Gamma_{nmkj}(t_1,t_2,t_3,t_4) \tag{9.5}$$

In other words, the asymptotic values of the fourth-order correlation functions of a stationary and narrowband field coincide with the corresponding fourth-order correlation functions of its equivalent Gaussian field.

The frequency-domain representation of $G_{nmkj}(t_1,t_2,t_3,t_4)$ is a straightforward generalization of the frequency-domain representation we introduced for the second-order correlation functions (Eq. (8.28)):

$$G_{nmkj}(\nu_1,\nu_2,\nu_3,\nu_4) = \int d^4t\, G_{nmkj}(t_1,t_2,t_3,t_4)\exp[2\pi i(\nu_1 t_1 + \nu_3 t_3 - \nu_2 t_2 - \nu_4 t_4)]$$

$$= \langle A_n(\nu_1)A_m^*(\nu_2)A_k(\nu_3)A_j^*(\nu_4)\rangle \tag{9.6}$$

We are interested here in the evaluation of (9.6) for a narrowband and stationary process. According to our definition of a stationary process,

$$G_{nmkj}(t_1,t_2,t_3,t_4) = G_{nmkj}(t_1 + \tau, t_2 + \tau, t_3 + \tau, t_4 + \tau) \tag{9.7}$$

for any τ. Let us consider $G_{nmkj}(t_1,t_2,t_3,t_4)$ for a moment as a function of $s = t_1 + t_2 + t_3 + t_4$ and three other independent linear combinations of t_1, t_2, t_3, and t_4. If we denote the value of s in the left-hand side as s_0, then its value on the right-hand side is $s_0 + 4\tau$. Since (9.7) holds for any τ, it follows that $G_{nmkj}(t_1,t_2,t_3,t_4)$ must be independent of $t_1 + t_2 + t_3 + t_4$. Let us derive the frequency-domain representation of an arbitrary function $F(t_1,t_2,t_3,t_4)$ of four arguments which is independent of their sum.

9.1.2 The Frequency-Domain Representation of a Function of Four Variables Which Is Independent of Their Sum

First, it is useful to introduce a new set of four variables, one of which being $t_1 + t_2 + t_3 + t_4$. Quite arbitrarily, we choose the following transformation:

$$s_1 = t_1 - t_2$$

$$s_2 = t_2 - t_3$$

$$s_3 = t_3 - t_4$$

$$s_4 = t_1 + t_2 + t_3 + t_4$$

The inverse transformation is

$$t_1 = \frac{1}{4}(3s_1 + 2s_2 + s_3 + s_4)$$

$$t_2 = \frac{1}{4}(-s_1 + 2s_2 + s_3 + s_4)$$

$$t_3 = \frac{1}{4}(-s_1 - 2s_2 + s_3 + s_4)$$

$$t_4 = \frac{1}{4}(-s_1 - 2s_2 - 3s_3 + s_4)$$

and its Jacobian is

$$\frac{\partial(t_1,t_2,t_3,t_4)}{\partial(s_1,s_2,s_3,s_4)} = \frac{1}{4}$$

Writing

$$W(s_1,s_2,s_3) = F(t_1,t_2,t_3,t_4)$$

we obtain the following expression for the frequency-domain representation of $F(t_1,t_2,t_3,t_4)$:

$$F(\nu_1,\nu_2,\nu_3,\nu_4) =$$

$$\frac{1}{4}\int d^4s W(s_1,s_2,s_3)\exp\{\tfrac{1}{2}\pi i[(3\nu_1 + \nu_2 - \nu_3 + \nu_4)s_1 + 2(\nu_1 - \nu_2 - \nu_3 + \nu_4)s_2$$

$$+ (\nu_1 - \nu_2 + \nu_3 + 3\nu_4)s_3 + (\nu_1 - \nu_2 + \nu_3 - \nu_4)s_4]\}$$

$$= W(\nu_1,\nu_1 - \nu_2,\nu_4)\delta(\nu_1 - \nu_2 + \nu_3 - \nu_4) \tag{9.8}$$

where

$$W(\mu_1,\mu_2,\mu_3) = \int d^3s W(s_1,s_2,s_3)\exp[2\pi i(s_1\mu_1 + s_2\mu_2 + s_3\mu_3)] \tag{9.9}$$

9.1.3 The Auxiliary Correlation Functions

After this brief technical detour, we return to our original problem, namely, the evaluation of the frequency-domain representation of $G_{nmkj}(t_1,t_2,t_3,t_4)$. Due to the independence of this function on $t_1 + t_2 + t_3 + t_4$, its frequency-domain representation will contain a factor $\delta(\nu_1 - \nu_2 + \nu_3 - \nu_4)$ as shown in (9.8). However, since $G_{nmkj}(t_1,t_2,t_3,t_4)$ does not generally vanish as the various differences of its time arguments tend to infinity, its frequency-domain representation will contain additional singularities. To expose these singularities, we introduce the auxiliary functions $\Delta G_{nmkj}(t_1,t_2,t_3,t_4)$:

$$\Delta G_{nmkj}(t_1,t_2,t_3,t_4) = G_{nmkj}(t_1,t_2,t_3,t_4) - \Gamma_{nmkj}(t_1,t_2,t_3,t_4) \tag{9.10}$$

By (9.5), these auxiliary functions tend to zero when any difference of their time arguments tends to infinity, and therefore their frequency-domain representation will not generally contain any singularities in addition to $\delta(\nu_1 - \nu_2 + \nu_3 - \nu_4)$. For a Gaussian field, the auxiliary correlation functions vanish, so they also may be regarded as a measure of the deviation of the given field from a pure Gaussian behavior. We introduce now the three-argument correlation functions

$$\mathrm{R}_{nmkj}(s_1,s_2,s_3) = \mathrm{G}_{nmkj}(t_1,t_2,t_3,t_4) \tag{9.11}$$

as well as the three-argument auxiliary functions

$$\Delta\mathrm{R}_{nmkj}(s_1,s_2,s_3) = \Delta\mathrm{G}_{nmkj}(t_1,t_2,t_3,t_4) \tag{9.12}$$

By (9.10) we have

$$\begin{aligned}\mathrm{R}_{nmkj}(s_1,s_2,s_3) &= \Delta\mathrm{R}_{nmkj}(s_1,s_2,s_3) + \mathrm{R}_{nj}(s_1+s_2+s_3)\mathrm{R}_{km}(-s_2)\\ &\quad + \mathrm{R}_{nm}(s_1)\mathrm{R}_{kj}(s_3)\end{aligned} \tag{9.13}$$

9.1.4 The Frequency-Domain Representation of the Fourth-Order Correlation Functions

The frequency-domain representation of $\mathrm{G}_{nmkj}(t_1,t_2,t_3,t_4)$ can now be derived by a repetitive application of (9.8) to the right-hand side of (9.13) term by term. Corresponding to the three terms in (9.13), we write $G_{nmkj}(\nu_1,\nu_2,\nu_3,\nu_4)$ as a sum of three terms:

$$G_{nmkj}(\nu_1,\nu_2,\nu_3,\nu_4) = G_1 + G_2 + G_3 \tag{9.14}$$

To compute G_1, we substitute $\Delta\mathrm{R}_{nmkj}(s_1,s_2,s_3)$ for $\mathrm{W}(s_1,s_2,s_3)$ in (9.9) and the resulting Fourier transform in (9.8). This yields

$$G_1 = \Delta R_{nmkj}(\nu_1,\nu_1-\nu_2,\nu_4)\delta(\nu_1-\nu_2+\nu_3-\nu_4) \tag{9.15}$$

To compute G_2, we substitute $\mathrm{R}_{nj}(s_1+s_2+s_3)\mathrm{R}_{km}(-s_2)$ for $\mathrm{W}(s_1,s_2,s_3)$ in (9.9). In this case,

$$W(\mu_1,\mu_2,\mu_3) = \int d^3s R_{nj}(s_1+s_2+s_3)\mathrm{R}_{km}(-s_2)\exp[2\pi i(s_1\mu_1+s_2\mu_2+s_3\mu_3)]$$

To compute this integral, we introduce a new set of variables u_1, u_2, u_3:

$$u_1 = s_1+s_2+s_3$$

$$u_2 = s_2$$

$$u_3 = s_3$$

In terms of these variables, we have

$$W(\mu_1,\mu_2,\mu_3) = \int d^3u R_{nj}(u_1) R_{km}(-u_2) \exp\{2\pi i[u_1\mu_1 + u_2(\mu_2 - \mu_1) + u_3(\mu_3 - \mu_1)]\}$$

$$= R_{nj}(\mu_1) R_{km}(\mu_1 - \mu_2)\delta(\mu_1 - \mu_3) \tag{9.16}$$

Using (9.8), we now obtain

$$G_2 = R_{nj}(\nu_1) R_{km}(\nu_2)\delta(\nu_1 - \nu_4)\delta(\nu_1 - \nu_2 + \nu_3 - \nu_4)$$

$$= R_{nj}(\nu_1) R_{km}(\nu_2)\delta(\nu_1 - \nu_4)\delta(\nu_2 - \nu_3) \tag{9.17}$$

To compute G_3, we substitute $R_{nm}(\nu_1)R_{kj}(\nu_3)$ for $W(s_1,s_2,s_3)$ in (9.9). This yields

$$W(\mu_1,\mu_2,\mu_3) = R_{nm}(\mu_1) R_{kj}(\mu_3)\delta(\mu_2) \tag{9.18}$$

Utilizing (9.8), we obtain

$$G_3 = R_{nm}(\nu_1) R_{kj}(\nu_3)\delta(\nu_1 - \nu_2)\delta(\nu_1 - \nu_2 + \nu_3 - \nu_4)$$

$$= R_{nm}(\nu_1) R_{kj}(\nu_3)\delta(\nu_1 - \nu_2)\delta(\nu_3 - \nu_4) \tag{9.19}$$

The frequency-domain representation of the fourth-order correlation functions can therefore be cast in the form

$$\begin{aligned} G_{nmkj}(\nu_1,\nu_2,\nu_3,\nu_4) &= \Delta R_{nmkj}(\nu_1,\nu_1 - \nu_2,\nu_4)\delta(\nu_1 - \nu_2 + \nu_3 - \nu_4) \\ &+ R_{nj}(\nu_1) R_{km}(\nu_3)\delta(\nu_1 - \nu_4)\delta(\nu_2 - \nu_3) \\ &+ R_{nm}(\nu_1) R_{kj}(\nu_3)\delta(\nu_1 - \nu_2)\delta(\nu_3 - \nu_4) \end{aligned} \tag{9.20}$$

The right-hand side is a sum of products of generally "well-behaved" functions and singular δ functions. The fundamental singularities of the frequency-domain representation are explicitly factored out in this form, making it convenient for analytic treatment. We will therefore refer to (9.20) as the *analytic representation* of the fourth-order correlation functions.

9.2 SPECIAL CASES

9.2.1 Polarized and Unpolarized Fields

Let us consider explicitly three common special cases: the cases of the polarized, the unpolarized, and the separable fields. In the case of the polarized field, there is only one nonvanishing component in the Jones vector, and the various quantities in (9.20) do not vanish only if all their indices are equal to the index of this nonvanishing component. In dealing with polarized fields, we will usually drop the indices, bearing in mind that all of them must be equal and have a given value. A similar situation prevails in the unpolarized case: here also all indices must have the same values; however, the common value may take each one of the two possible values, 1 or 2. Furthermore, in the unpolarized case the various quantities in (9.20) are independent of this common value of the indices. Therefore, in the unpolarized case also there is no point in keeping the indices. To sum up, in the cases of the polarized and the unpolarized fields we may use a simplified version of (9.20), namely,

$$\begin{aligned} G(\nu_1,\nu_2,\nu_3,\nu_4) &= \Delta R(\nu_1,\nu_1-\nu_2,\nu_4)\delta(\nu_1-\nu_2+\nu_3-\nu_4) \\ &\quad + R(\nu_1)R(\nu_3)\delta(\nu_1-\nu_4)\delta(\nu_2-\nu_3) \\ &\quad + R(\nu_1)R(\nu_3)\delta(\nu_1-\nu_2)\delta(\nu_3-\nu_4) \end{aligned} \tag{9.21}$$

It must be borne in mind, though, that the interpretations of this equation for the two cases is slightly different: in the case of the polarized field it represents the statistics of the single nonvanishing field component, and in the case of the unpolarized field it represents the *common* statistics of the two field components.

9.2.2 Separable Field

We turn now to the more general case of the separable field. Since in the separable field the two polarization modes are uncorrelated, it follows that for such a field $G_{nmkj}(t_1,t_2,t_3,t_4)$ does not vanish only if the four Jones vector components it comprises break into two pairs, each consisting of two components with equal indices, one being a regular value and the other a complex conjugate (including the case in which all four components have equal indices). Thus, the general form of the fourth-order correlation functions in the case of the separable field is

$$
\begin{aligned}
&G_{nmkj}(t_1,t_2,t_3,t_4) = \\
&\qquad G_{nnnn}(t_1,t_2,t_3,t_4)\delta_{nm}\delta_{mk}\delta_{kj} \\
&\qquad + [G_{nn}(t_1,t_2)G_{kk}(t_3,t_4)\delta_{nm}\delta_{kj} + G_{nn}(t_1,t_4)G_{kk}(t_3,t_2)\delta_{nj}\delta_{mk}](1 - \delta_{nk})
\end{aligned}
\tag{9.22}
$$

The first term describes the case in which all indices are equal, and the second term describes the two possibilities in which the equal-index pairs have different indices. Furthermore, the fourth-order correlation functions of the equivalent Gaussian field of a separable source are

$$
\Gamma_{nmkj}(t_1,t_2,t_3,t_4) = G_{nn}(t_1,t_2)G_{kk}(t_3,t_4)\delta_{nm}\delta_{kj} + G_{nn}(t_1,t_4)G_{kk}(t_3,t_2)\delta_{nj}\delta_{mk} \tag{9.23}
$$

From (9.10) we find that, in this case,

$$
\begin{aligned}
\Delta G_{nmkj}(t_1,t_2,t_3,t_4) &= [G_{nnnn}(t_1,t_2,t_3,t_4) - G_{nn}(t_1,t_2)G_{nn}(t_3,t_4) \\
&\quad - G_{nn}(t_1,t_4)G_{nn}(t_3,t_2)]\delta_{nm}\delta_{mk}\delta_{kj}
\end{aligned}
\tag{9.24}
$$

Consequently, the frequency-domain representation of the fourth-order correlation functions of a separable field can be cast in the form

$$
\begin{aligned}
G_{nmkj}(\nu_1,\nu_2,\nu_3,\nu_4) &= \Delta R_{nnnn}(\nu_1,\nu_1 - \nu_2,\nu_4)\delta(\nu_1 - \nu_2 + \nu_3 - \nu_4)\delta_{nm}\delta_{mk}\delta_{kj} \\
&\quad + R_{nn}(\nu_1)R_{kk}(\nu_3)\delta(\nu_1 - \nu_4)\delta(\nu_2 - \nu_3)\delta_{nj}\delta_{mk} \\
&\quad + R_{nn}(\nu_1)R_{kk}(\nu_3)\delta(\nu_1 - \nu_2)\delta(\nu_3 - \nu_4)\delta_{nm}\delta_{kj}
\end{aligned}
\tag{9.25}
$$

where $\Delta R_{nnnn}(\mu_1,\mu_2,\mu_3)$ is the Fourier transform of

$$
\Delta R_{nnnn}(s_1,s_2,s_3) = R_{nnnn}(s_1,s_2,s_3) - R_{nn}(s_1)R_{nn}(s_3) - R_{nn}(s_1 + s_2 + s_3)R_{nn}(-s_2)
$$

We see that in the separable field case, $G_{nmkj}(\nu_1,\nu_2,\nu_3,\nu_4)$ is determined by only four independent functions: the two functions $\Delta R_{nnnn}(\nu_1,\nu_2,\nu_3)$ and the two functions $R_{nn}(\nu)$. Each of these functions is associated with only one of the polarization modes of the field.

9.2.3 Gaussian Field

Finally, we note that for a Gaussian field

$$G_{nmkj}(\nu_1,\nu_2,\nu_3,\nu_4) = R_{nj}(\nu_1)R_{km}(\nu_3)\delta(\nu_1-\nu_4)\delta(\nu_2-\nu_3) + R_{nm}(\nu_1)R_{kj}(\nu_3)\delta(\nu_1-\nu_2)\delta(\nu_3-\nu_4) \quad (9.26)$$

since, in this case, $\Delta R_{nmkj}(\nu_1,\nu_3,\nu_3)$ vanishes.

9.3 THE RANDOM-PHASE FIELD

In Section 8.4 we derived the second-order correlation function of the random-phase field. Here we will derive the fourth-order correlation functions of this field. As we have shown, the random-phase field is separable. For simplicity, we treat here only one of its polarization components, and drop the polarization mode subscript. Using (8.50), we write the fourth-order correlation function of the random-phase field as

$$G(t_1,t_2,t_3,t_4) = \langle I\rangle^2 \exp[-2\pi i\nu_0(t_1-t_2+t_3-t_4)] \cdot \langle \exp\{i[\phi(t_1)-\phi(t_2)+\phi(t_3)-\phi(t_4)]\}\rangle \quad (9.27)$$

The average appearing in (9.27) can be computed with the help of (8.56):

$$\langle \exp\{i[\phi(t_1)-\phi(t_2)+\phi(t_3)-\phi(t_4)]\}\rangle = \exp\{-\tfrac{1}{2}\langle[\phi(t_1)-\phi(t_2)+\phi(t_3)-\phi(t_4)]^2\rangle\}$$

Writing

$$G_\phi(t_1,t_2) = \langle\phi(t_1)\phi(t_2)\rangle$$

we obtain

$$\langle[\phi(t_1)-\phi(t_2)+\phi(t_3)-\phi(t_4)]^2\rangle = \sum_{j,k=1}^{4} (-1)^{j+k} G_\phi(t_j,t_k) \quad (9.28)$$

We can express the phase correlation function $G_\phi(t_1,t_2)$ in terms of $\Lambda(t_1-t_2)$ (8.60). Indeed,

$$G_\phi(t_1,t_2) = <[2\pi\int_0^{t_1} dt\Delta\nu(t) + \phi_0][2\pi\int_0^{t_2} dt\Delta\nu(t) + \phi_0)]>$$

$$= 4\pi^2\int_0^{t_1}\int_0^{t_2} dtdt'<\Delta\nu(t)\Delta\nu(t')> + <\phi_0^2>$$

$$= 4\pi^2\int_0^{t_1}\int_0^{t_2} dtdt' G_\nu(t - t') + <\phi_0^2> \tag{9.29}$$

The region of integration in the $t\times t'$ plane for this integral is shown in Figure 9.1.

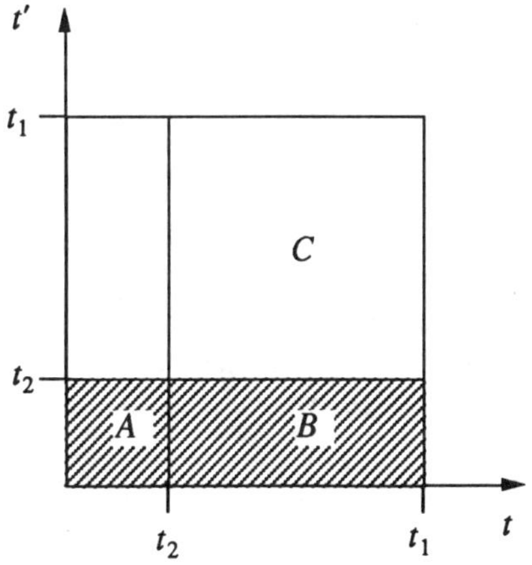

Figure 9.1 The integration regions in the $t\times t'$ plane. The hatched region denotes the integration region in (9.29).

Let us denote the integrals over the regions A, B, and C with I_A, I_B, and I_C respectively. From (8.60) we have

$$I_A = \Lambda(t_2)$$

$$I_C = \Lambda(t_1 - t_2)$$

$$I_A + 2I_B + I_C = \Lambda(t_1)$$

It follows therefore that

$$G_\phi(t_1,t_2) = I_A + I_B = \frac{1}{2}[\Lambda(t_1) + \Lambda(t_2) - \Lambda(t_1 - t_2)] + < \phi_0^2> \tag{9.30}$$

Using this result on the right-hand side of (9.28), we obtain

$$<[\phi(t_1) - \phi(t_2) + \phi(t_3) - \phi(t_4)]^2> = \frac{1}{2}\sum_{j,k}(-1)^{j+k}[\Lambda(t_j) + \Lambda(t_k) - \Lambda(t_j - t_k) + 2<\phi_0^2>]$$
$$= -\frac{1}{2}\sum_{j,k}(-1)^{j+k}\Lambda(t_j - t_k)$$

Since $\Lambda(t)$ is an even function, this sum may be ordered, yielding

$$<[\phi(t_1) - \phi(t_2) + \phi(t_3) - \phi(t_4)]^2> = -\sum_{j=1}^{3}\sum_{k=j+1}^{4}(-1)^{k+j}\Lambda(t_j - t_k) \tag{9.31}$$

We are now ready to present the explicit expression for the fourth-order correlation functions of the random-phase field:

$$G(t_1,t_2,t_3,t_4)$$
$$= <I>^2\exp\left[-2\pi i\nu_0(t_1 - t_2 + t_3 - t_4) + \frac{1}{2}\sum_{j=1}^{3}\sum_{k=j+1}^{4}(-1)^{k+j}\Lambda(t_j - t_k)\right] \tag{9.32}$$

As expected, this function depends only on the differences of the various time arguments. This expression is a special case of a general formula derived by Picinbono and Boileau [3]. The three-argument correlation function is given by

$$R(s_1,s_2,s_3) = <I>^2\exp[-2\pi i\nu_0(s_1 + s_3)]$$
$$\cdot\exp\{-\tfrac{1}{2}[\Lambda(s_1) + \Lambda(s_2) + \Lambda(s_3)$$
$$-\Lambda(s_1 + s_2) - \Lambda(s_2 + s_3) + \Lambda(s_1 + s_2 + s_3)]\} \tag{9.33}$$

We turn now to the computation of the corresponding correlation functions of the equivalent Gaussian field. Using (9.3) and (8.70), we obtain

$$\Gamma(t_1,t_2,t_3,t_4) = <I>^2\exp[-2\pi i\nu_0(t_1 - t_2 + t_3 - t_4)]$$
$$\cdot\left\{\exp\{-\tfrac{1}{2}[\Lambda(t_1 - t_2) + \Lambda(t_3 - t_4)]\}\right.$$
$$\left.+\exp\{-\tfrac{1}{2}[\Lambda(t_1 - t_4) + \Lambda(t_3 - t_2)]\}\right\} \tag{9.34}$$

or, in terms of the s variables,

$$\Gamma(t_1,t_2,t_3,t_4) = \langle I\rangle^2 \exp[-2\pi i \nu_0(s_1+s_3)] \cdot \left\{ \exp\{-\tfrac{1}{2}[\Lambda(s_1)+\Lambda(s_3)]\} + \exp\{-\tfrac{1}{2}[\Lambda(s_1+s_2+s_3)+\Lambda(s_2)]\} \right\}$$

The auxiliary fourth-order correlation function can therefore be written as

$$\begin{aligned}\Delta R(s_1,s_2,s_3) &= \langle I\rangle^2 \exp\{2\pi i \nu_0(s_1+s_3) - \tfrac{1}{2}[\Lambda(s_1)+\Lambda(s_2)+\Lambda(s_3)+\Lambda(s_1+s_2+s_3)]\}\\ &\cdot \left\{ \exp\{\tfrac{1}{2}[\Lambda(s_1+s_2)+\Lambda(s_2+s_3)]\} - \exp\{\tfrac{1}{2}[\Lambda(s_2)+\Lambda(s_1+s_2+s_3)]\}\right.\\ &\left. - \exp\{\tfrac{1}{2}[\Lambda(s_1)+\Lambda(s_3)]\} \right\}\end{aligned} \tag{9.35}$$

We remind the reader that the auxiliary correlation function $\Delta R(s_1,s_2,s_3)$ vanishes whenever the absolute value of any of its arguments tends to infinity. This property can be verified explicitly in (9.35) using the asymptotic relation (8.69).

9.4 THE OPTICAL INTENSITY POWER SPECTRUM

9.4.1 The Time-Domain Representation of the Optical Intensity Power Spectrum

The correlation function of the optical intensity $I(t)$ is given by

$$\begin{aligned}\langle I(t+\tau)I(t)\rangle &= \langle \mathrm{Tr}[\mathbf{A}(t+\tau)\mathbf{A}^\dagger(t+\tau)]\mathrm{Tr}[\mathbf{A}(t)\mathbf{A}^\dagger(t)]\rangle\\ &= \sum_{j,k} G_{jjkk}(t+\tau,t+\tau,t,t)\end{aligned} \tag{9.36}$$

The functions $G_{jjkk}(t+\tau,t+\tau,t,t)$ can be expressed in terms of the three-argument correlation functions defined in (9.11) as follows:

$$G_{jjkk}(t+\tau,t+\tau,t,t) = R_{jjkk}(0,\tau,0) \tag{9.37}$$

and therefore

$$\langle I(t+\tau)I(t)\rangle = \sum_{j,k} R_{jjkk}(0,\tau,0) \tag{9.38}$$

The power spectrum of the optical intensity, which will be denoted by $S(f)$, is therefore given by

$$S(f) = \sum_{j,k} \int d\tau e^{2\pi i f \tau} R_{jjkk}(0,\tau,0) \tag{9.39}$$

We denote the argument of the power spectrum of the optical intensity with f rather than ν, to allude to the fact that the range of values of f for which $S(f)$ is significantly nonzero overlaps the radio-frequency band rather than the optical-frequency band. By (9.13),

$$R_{jjkk}(0,\tau,0) = \Delta R_{jjkk}(0,\tau,0) + R_{jk}(\tau)R_{kj}(-\tau) + R_{jj}(0)R_{kk}(0) \tag{9.40}$$

We therefore write $S(f)$ as a sum of three terms:

$$S(f) = S_1(f) + S_2(f) + S_3(f) \tag{9.41}$$

where

$$S_1(f) = \sum_{j,k} \int d\tau\, e^{2\pi i f \tau} \Delta R_{jjkk}(0,\tau,0) \tag{9.42}$$

$$S_2(f) = \sum_{j,k} \int d\tau e^{2\pi i f \tau} R_{jk}(\tau) R_{kj}(-\tau) \tag{9.43}$$

$$S_3(f) = \sum_{j,k} R_{jj}(0) R_{kk}(0) \delta(f) = \langle I \rangle^2 \delta(f) \tag{9.44}$$

The second term can also be written using matrix notation:

$$S_2(f) = \int d\tau\, e^{2\pi i f \tau} \mathrm{Tr}[\mathbf{R}(\tau)\mathbf{R}(-\tau)] \tag{9.45}$$

It is easy to verify that all four functions $S(f)$, $S_1(f)$, $S_2(f)$, $S_3(f)$ are even. To sum up, the time-domain representation of the intensity power spectrum reads

$$S(f) = \int d\tau e^{2\pi i f \tau} \{ \sum_{j,k} \Delta R_{jjkk}(0,\tau,0) + \mathrm{Tr}[\mathbf{R}(\tau)\mathbf{R}(-\tau)] \} + \langle I \rangle^2 \delta(f) \tag{9.46}$$

9.4.2 The Intensity Noise Power Spectrum and the Intensity Variance

In many cases, it is more convenient to deal with the intensity *noise* power spectrum $N(f)$ rather than with the intensity power spectrum. The intensity noise power spectrum is the Fourier transform of the intensity *covariance* function $C_I(\tau)$:

$$C_I(\tau) = <[I(t+\tau) - <I>][I(t) - <I>]> = <I(t+\tau)I(t)> - <I>^2$$

i.e.,

$$N(f) = \int d\tau e^{2\pi i f \tau} C_I(\tau)$$

The functions $S(f)$ and $N(f)$ are related as follows:

$$S(f) = N(f) + <I>^2 \delta(f) \tag{9.47}$$

The relation between $S(f)$ and $N(f)$ is portrayed schematically in Figure 9.2.

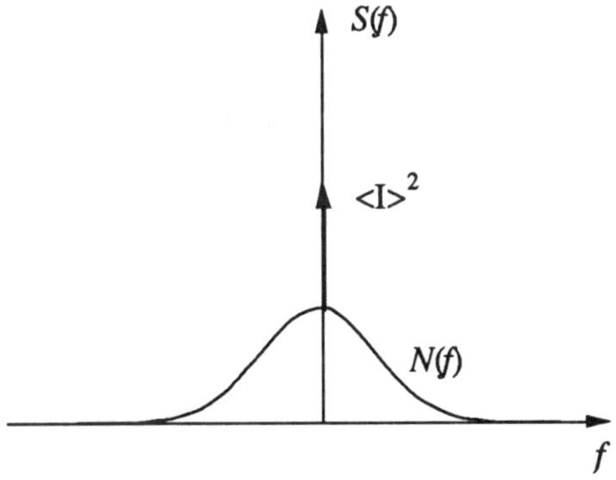

Figure 9.2 The optical intensity and optical intensity noise power spectra.

The *intensity variance*, Var[I], is equal to the intensity covariance function at $\tau = 0$:

$$\text{Var}[I] = <I^2> - <I>^2 = C_I(0) \tag{9.48}$$

From (9.46), it follows that the intensity variance also equals the total noise power; i.e.,

$$\text{Var}[I] = \int df\, N(f) \tag{9.49}$$

We can express Var[I] in terms of the second and fourth-order correlation functions:

$$\text{Var}[I] = \sum_{j,k} \Delta R_{jjkk}(0,0,0) + \text{Tr}[\mathbf{R}^2(0)]$$

In the Gaussian case, since ΔR_{jjkk} vanishes, we have

$$\mathrm{Var}[I] = \mathrm{Tr}[\mathbf{R}^2(0)] \tag{9.50}$$

It is interesting to note that in the case of a polarized Gaussian field, we have, by (8.34),

$$\mathrm{Var}[I] = \langle I \rangle^2 \tag{9.51}$$

9.4.3 The Frequency-Domain Representation

To obtain the frequency-domain representation of the field intensity power spectrum $S(f)$, we take the Fourier transform of (9.36):

$$S(f) = \int d\tau e^{2\pi i f\tau} \langle \mathrm{Tr}[\mathbf{A}(t+\tau)\mathbf{A}^\dagger(t+\tau)]\mathrm{Tr}[\mathbf{A}(t)\mathbf{A}^\dagger(t)] \rangle$$

We express now the time-domain Jones vectors in terms of their frequency-domain counterparts:

$$S(f) = \int d\tau d^4\nu \exp\{2\pi i[f\tau - \nu_1(t+\tau) + \nu_2(t+\tau) - \nu_3 t + \nu_4 t]\}$$
$$\cdot \langle \mathrm{Tr}[A(\nu_1)A^\dagger(\nu_2)]\mathrm{Tr}[A(\nu_3)A^\dagger(\nu_4)] \rangle$$

As we have seen, the integrand contains a factor $\delta(\nu_1 - \nu_2 + \nu_3 - \nu_4)$ (not shown explicitly). Therefore, we can also write

$$S(f) = \int d\tau d^4\nu \exp[2\pi i\tau(f - \nu_1 + \nu_2)]$$
$$\cdot \langle \mathrm{Tr}[A(\nu_1)A^\dagger(\nu_2)]\mathrm{Tr}[A(\nu_3)A^\dagger(\nu_4)] \rangle$$

Performing the integration over τ, we obtain

$$S(f) = \int d^4\nu \langle \mathrm{Tr}[A(\nu_1)A^\dagger(\nu_2)]\mathrm{Tr}[A(\nu_3)A^\dagger(\nu_4)] \rangle \delta(f - \nu_1 + \nu_2) \tag{9.52}$$

The right-hand side can also be expressed in terms of the fourth-order field correlation functions:

$$S(f) = \sum_{j,k} \int d^4 \nu G_{jjkk}(\nu_1, \nu_2, \nu_3, \nu_4) \delta(f - \nu_1 + \nu_2) \tag{9.53}$$

We now substitute for the correlation functions their analytic representation (Eq. (9.20)). Breaking up the resulting expression, we obtain the three terms $S_1(f)$, $S_2(f)$, and $S_3(f)$. For $S_1(f)$ we obtain

$$\begin{aligned} S_1(f) &= \sum_{j,k} \int d^4 \nu \Delta R_{jjkk}(\nu_1, \nu_1 - \nu_2, \nu_4) \delta(\nu_1 - \nu_2 + \nu_3 - \nu_4) \delta(f - \nu_1 + \nu_2) \\ &= \sum_{j,k} \int d^2 \nu \Delta R_{jjkk}(\nu_1, f, \nu_2) \end{aligned}$$

Regarding $S_2(f)$, we obtain

$$\begin{aligned} S_2(f) &= \sum_{j,k} \int d^4 \nu R_{jk}(\nu_1) R_{kj}(\nu_3) \delta(\nu_1 - \nu_4) \delta(\nu_2 - \nu_3) \delta(f - \nu_1 + \nu_2) \\ &= \int d\nu \mathrm{Tr}[\boldsymbol{R}(\nu) \mathbf{R}(\nu + f)] \end{aligned}$$

and finally,

$$\begin{aligned} S_3(f) &= \sum_{j,k} \int d^4 \nu R_{jj}(\nu_1) R_{kk}(\nu_3) \delta(\nu_1 - \nu_2) \delta(\nu_3 - \nu_4) \delta(f - \nu_1 + \nu_2) \\ &= \left[\int d\nu \mathrm{Tr} \boldsymbol{R}(\nu) \right]^2 \delta(f) \end{aligned}$$

To sum up, the expression of $S(f)$ in terms of frequency-domain functions can be written in the form

$$S(f) = \sum_{j,k} \int d^2 \nu \Delta R_{jjkk}(\nu_1, f, \nu_2) + \int d\nu \mathrm{Tr}[\boldsymbol{R}(\nu) \boldsymbol{R}(\nu + f)] + \langle I \rangle^2 \delta(f) \tag{9.54}$$

This is the frequency-domain counterpart of (9.46). We invite the reader to check directly that (9.54) and (9.46) are completely equivalent. In fact, a direct derivation of (9.54) from (9.46) is simpler than the derivation that we presented here. We have chosen to present here a derivation based on the analytic representation (9.20), since the same approach will serve us in two more instances later: in the derivation of the output intensity power spectrum of time-independent and time-periodic networks. The second term in (9.54) and (9.46) represents the intensity power spectrum of the equivalent Gaussian field, and may be regarded as the *Gaussian component* of the spectrum.

Since the fourth-order correlation functions of the field are not generally available, the expressions that we have derived here for the computation of $S(f)$ have a limited practical value, with the exception of the Gaussian field case, in which the auxiliary fourth-order correlation function vanishes. However, we will be mainly concerned not with the evaluation of $S(f)$ itself, but rather with the problem of how the intensity noise spectrum is affected by a linear network. The usefulness of (9.54) will become apparent when we will tackle this problem in the next chapter.

9.4.4 Special Cases

Let us consider the case of the separable field. According to (9.25), we can write

$$S(f) = \sum_n N_n(f) + \langle I \rangle^2 \delta(f) \tag{9.55}$$

where

$$N_n(f) = \int d^2\nu \Delta R_{nnnn}(\nu_1, f, \nu_2) + \int d\nu P_n(\nu) P_n(\nu + f) \tag{9.56}$$

may be regarded as the intensity noise power spectrum associated with the nth field polarization mode.

In the Gaussian field case, since the auxiliary correlation function vanishes, we obtain [2]

$$N(f) = \int d\nu \mathrm{Tr}[\boldsymbol{R}(\nu)\boldsymbol{R}(\nu + f)] \tag{9.57}$$

It is instructive to consider the case of the random-phase field. According to (8.50),

$$\langle I(t+\tau) I(t) \rangle = |a|^4 = \langle I \rangle^2$$

so that, in this case,

$$S(f) = \langle I \rangle^2 \delta(f)$$

Comparing this to (9.54), we conclude that for each of the polarization modes of a random-phase field,

$$\int d^2\nu \Delta R(\nu_1, f, \nu_2) + \int d\nu R(\nu) R(\nu + f) = 0 \tag{9.58}$$

The time domain version of this equation is

$$\Delta R(0,\tau,0) + R(\tau)R(-\tau) = 0 \tag{9.59}$$

This relation can also be verified explicitly. From (9.35) we obtain

$$\Delta R(0,\tau,0) = \langle I\rangle^2 e^{-\Lambda(\tau)}[e^{\Lambda(\tau)} - e^{\Lambda(\tau)} - 1] = -\langle I\rangle^2 e^{-\Lambda(\tau)}$$

Recalling (8.70),

$$R(\tau) = \langle I\rangle \exp[-\tfrac{1}{2}\Lambda(\tau)]$$

and because $\Lambda(\tau)$ is an even function, we obtain (9.59).

9.5 THE OPTICAL INTENSITY POWER SPECTRUM OF A REAL FIELD AND ITS ASSOCIATED COMPLEX FIELD

In order to allow the use of complex Jones vectors for the evaluation of $S(f)$, we must provide the relation between the intensity power spectrum of a real field and the intensity power spectrum of its associated complex field. We start by noting that the field intensity correlation function can be expressed as an average of two scalar products:

$$\langle I_A(t+\tau)I_A(t)\rangle = \langle(\mathbf{A}^*(t+\tau)\cdot\mathbf{A}(t+\tau))(\mathbf{A}^*(t)\cdot\mathbf{A}(t))\rangle \tag{9.60}$$

The general representation of a narrowband real Jones vector $\mathbf{A}(t)$ was given by (8.42):

$$\mathbf{A}(t) = \mathbf{a}_1(t)\cos[2\pi\nu(t-\theta)] + \mathbf{a}_2(t)\sin[2\pi\nu(t-\theta)]$$

To reduce the size of the expressions that are to be presented, we introduce the notations

$$c = \cos[2\pi\nu(t-\theta)]$$

$$s = \sin[2\pi\nu(t-\theta)]$$

$$c_\tau = \cos[2\pi\nu(t+\tau-\theta)]$$

$$s_\tau = \sin[2\pi\nu(t + \tau - \theta)]$$

$$\mathbf{a}_n = \mathbf{a}_n(t)$$

$$a_{n\tau} = \mathbf{a}_n(t + \tau)$$

Using these notations, we express the correlation function of the field intensity as follows:

$$\begin{aligned}<\mathrm{I}_A(t+\tau)\mathrm{I}_A(t)> &= <[\mathbf{a}_{1\tau}{}^2 c_\tau^2 + \mathbf{a}_{2\tau}{}^2 s_\tau^2 + 2(\mathbf{a}_{1\tau}\cdot \mathbf{a}_{2\tau})s_\tau c_\tau] \\ &\cdot [\mathbf{a}_1{}^2 c^2 + \mathbf{a}_2{}^2 s^2 + 2(\mathbf{a}_1 \cdot \mathbf{a}_2)sc]>\end{aligned}$$

Since the random variable θ is independent of the random vectors $\mathbf{a}_1(t)$ and $\mathbf{a}_2(t)$, we obtain

$$\begin{aligned}<\mathrm{I}_A(t+\tau)\mathrm{I}_A(t)> = &<\mathbf{a}_{1\tau}{}^2\mathbf{a}_1{}^2><c_\tau^2c^2> + <\mathbf{a}_{1\tau}{}^2\mathbf{a}_2{}^2><c_\tau^2s^2> + 2<\mathbf{a}_{1\tau}{}^2(\mathbf{a}_1\cdot \mathbf{a}_2)><c_\tau^2sc> \\ &+ <\mathbf{a}_{2\tau}{}^2\mathbf{a}_1{}^2><s_\tau^2c^2> + <\mathbf{a}_{2\tau}{}^2\mathbf{a}_2{}^2><s_\tau^2s^2> + <\mathbf{a}_{2\tau}{}^2(\mathbf{a}_1\cdot \mathbf{a}_2)><s_\tau^2sc> \\ &+ 2<(\mathbf{a}_{1\tau}\cdot \mathbf{a}_{2\tau})\mathbf{a}_1{}^2><s_\tau c_\tau c^2> + 2<(\mathbf{a}_{1\tau}\cdot \mathbf{a}_{2\tau})\mathbf{a}_2{}^2><s_\tau c_\tau s^2> \\ &+ 4<(\mathbf{a}_{1\tau}\cdot \mathbf{a}_{2\tau})(\mathbf{a}_1\cdot \mathbf{a}_2)><s_\tau c_\tau sc>\end{aligned} \tag{9.61}$$

The trigonometric functions averages are given by

$$\begin{aligned}<c_\tau^2c^2> &= \frac{1}{T}\int_0^T d\theta \cos^2[2\pi\nu(t+\tau-\theta)]\cos^2(t-\theta) \\ &= \frac{1}{4} + \frac{1}{8}\cos(4\pi\nu\tau) \\ <s_\tau^2s^2> &= \frac{1}{4} + \frac{1}{8}\cos(4\pi\nu\tau) \\ <c_\tau^2s^2> &= \frac{1}{4} - \frac{1}{8}\cos(4\pi\nu\tau) \\ <s_\tau^2c^2> &= <c_\tau^2s^2> \\ <s_\tau c_\tau sc> &= \frac{1}{8}\cos(4\pi\nu\tau) \\ <s_\tau c_\tau c^2> &= \frac{1}{8}\sin(4\pi\nu\tau) \\ <s_\tau c_\tau s^2> &= -\frac{1}{8}\sin(4\pi\nu\tau)\end{aligned}$$

$<scc_\tau^2> \quad = <s_\tau c_\tau c^2>$

$<scs_\tau^2> \quad = <s_\tau c_\tau s^2>$

When the Fourier transform of (9.61) is performed for the computation of $S(f)$, it is found that the terms that contain the factors $\cos(4\pi\nu\tau)$ or $\sin(4\pi\nu\tau)$ give rise to spectral components centered at the frequencies 2ν and -2ν, while terms that do not contain trigonometric functions create a spectral component centered at the origin of the frequency axis, as shown schematically in Figure 9.3.

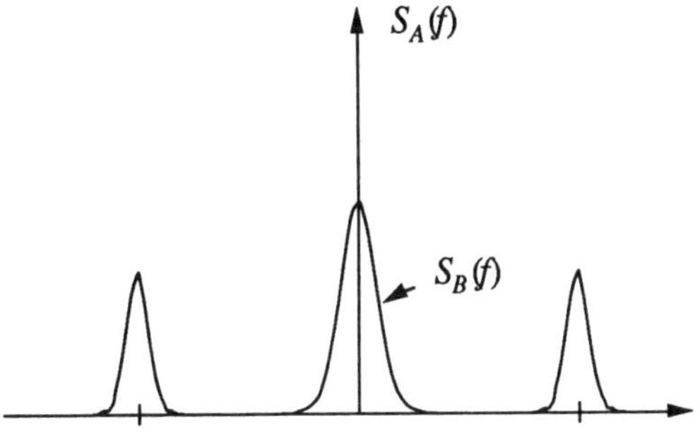

Figure 9.3 The three spectral features in the intensity power spectrum $S_A(f)$ of the real process. Only the central component, which is equal to the intensity power spectrum $S_B(f)$ of the associated complex process, has practical relevance.

Since no optical detector can respond to signals fluctuating with an optical frequency, only the central spectral component has a practical significance. Regarding the detector as a linear low-pass filter for the intensity with a cutoff frequency much smaller than ν, we refer to the central component as the filtered outcome of the intensity fluctuations. Collecting the terms in (9.61) that contribute to this component, we obtain

$$<I_A(t)I_A(t+\tau)>|_{\text{filtered}} = \frac{1}{4}(<a_{1\tau}^2 a_1^2> + <a_{1\tau}^2 a_2^2> + <a_{2\tau}^2 a_1^2> + <a_{2\tau}^2 a_2^2>) \qquad (9.62)$$

Let us consider now the corresponding intensity correlation function of the associated complex process $\mathbf{B}(t)$ (Eq. (8.43)):

$$\mathbf{B}(t) = \frac{1}{\sqrt{2}}[\mathbf{a}_1(t) + \mathbf{a}_2(t)]e^{-2\pi i\nu(t-\theta)}$$

Applying (9.59) to $\mathbf{B}(t)$ we obtain

$$\begin{aligned}
<I_B(t)I_B(t+\tau)> &= <(\mathbf{B}^*(t+\tau)\cdot \mathbf{B}(t+\tau))(\mathbf{B}^*(t)\cdot \mathbf{B}(t))> \\
&= \frac{1}{4}<(\mathbf{a}_{1\tau}^2 + \mathbf{a}_{2\tau}^2)(\mathbf{a}_1^2 + \mathbf{a}_2^2)> \\
&= <I_A(t+\tau)I_A(t)>|_{\text{filtered}}
\end{aligned} \tag{9.63}$$

so that

$$S_A(f)|_{\text{filtered}} = S_B(f) \tag{9.64}$$

We see that the relevant component of the intensity power spectrum of the real process is identical to the intensity power spectrum of the associated complex process. This justifies the use of the complex representation for the computation of the intensity power spectrum of a real process.

9.6 THE RELATIVE INTENSITY NOISE POWER SPECTRUM

Since the intensity noise power spectrum $N(f)$ is generally proportional to the second power of the average intensity $<I>^2$, it is convenient to introduce a normalized noise power spectrum, which has been called in the literature *relative intensity noise power spectrum*, and denoted by $RIN(f)$:

$$RIN(f) = \frac{N(f)}{<I>^2} \tag{9.65}$$

In practice, it often occurs that f_c is much larger than any relevant frequency scale (like, for instance, in specifying a laser diode with f_c = 10 GHz and electrical bandwidth of 100 MHz). In these cases the frequency dependence of $RIN(f)$ can be neglected, and it is enough to specify its value at $f = 0$ only. Thus, whenever *RIN* without a frequency argument is specified, it should be interpreted as the value of the relative intensity noise power spectrum at $f = 0$.

In technical data sheets, *RIN* is given in units of dB/Hz. A typical value of *RIN* for laser diodes lies in the range between –100 dB/Hz to –150 dB/Hz. To compute the total relative intensity noise contained in a given bandwidth B, it is necessary first to convert *RIN* to its absolute numerical value and then to multiply this numerical value by B. For example, a laser diode with a specification *RIN* = –100 dB/Hz, will exhibit, in a bandwidth of 100 KHz, a total relative intensity noise of 10^{-5}. This number also represents the relative variance of the filtered detector current.

Let us evaluate $RIN(f)$ for a polarized Gaussian field with a Gaussian lineshape; i.e.,

$$R(\nu) = \frac{\langle I \rangle}{\sqrt{2\pi f_c^2}} \exp\left[-\frac{(\nu-\nu_0)^2}{2f_c^2}\right]$$

Using (9.54) and recalling that the auxiliary correlation function vanishes for a Gaussian field, we find for this case

$$RIN(f) = \frac{1}{\sqrt{4\pi f_c^2}} \exp\left[-\frac{f^2}{4f_c^2}\right] \tag{9.66}$$

We see, in this case, that

$$RIN \approx 1/f_c = \tau_c \tag{9.67}$$

In general, it is not possible to estimate the magnitude of *RIN* from the optical lineshape only. In particular, the value of *RIN* for the field of highly coherent and amplitude-stabilized laser diodes may be significantly lower than the estimate (9.67). This is due to the fact that phase fluctuations of the field incident on a detector do not show up in its output current. We will see, however, in the next section, that optical networks may convert phase fluctuations in the input field to amplitude fluctuations in the output field. Therefore, even if the value of *RIN* in the network input field is significantly lower than $1/f_c$, it may increase to $1/f_c$ in the output field.

Consider, for instance, a highly coherent laser diode with f_c = 100 MHz and *RIN* = –150 dB/Hz. When this laser diode is coupled to an optical network, the value of *RIN* in the output field may be as high as 10^{-8} Hz^{-1}, or –80 dB/Hz. This increase in the relative intensity noise is due to conversion of phase fluctuations in the input field to amplitude fluctuations in the output field. Examples of this phase-to-amplitude noise conversion will be considered in detail in the next chapter. Generally, if there is no information on the field other than its coherence time, (9.67) may be used to estimate *RIN*.

9.7 A COMPARISON BETWEEN FIELD-INDUCED NOISE AND SHOT NOISE

As we already mentioned, the generation of noise by the statistical field fluctuations is only one of the many possible noise mechanisms that may be present in a given application. There is, however, one fundamental noise present in any optical quantum detection system: the shot noise [4]. It is therefore of interest to compare the magnitude of shot noise to the magnitude of field-induced noise. We would like to point out that in photoconductive detectors, it is customary to add together the shot noise and the generation-recombination noise powers [4], resulting in an equivalent noise with a magnitude twice as large as the shot noise.

The power spectrum $N^{(S)}_i$ of shot noise in the detector output current is given by [4]

$$N^{(S)}_i = q\eta\langle I\rangle$$

where q is the electron charge, with a value of 1.6×10^{-19} C, η is the detector responsivity measured in A/W, and $\langle I\rangle$ is the average incident optical power measured in watts. We would like to remind the reader that we use here the two-sided spectral representation; i.e., we extend the spectrum in both the positive and the negative frequency axis. In texts using the one-sided spectral representation, there is, accordingly, an extra factor of 2 on the right-hand side of the above equation.

The shot noise power spectrum in the detector current is measured in units of A^2/Hz, which is the appropriate unit for a current noise power spectrum. In order to compare this noise to the field-induced noise, we have to convert its units to W^2/Hz, or, in other words, to define a *shot noise equivalent* optical intensity noise $N^{(S)}_I$. This is done simply by dividing $N^{(S)}_i$ by η^2, resulting in

$$N^{(S)}_I = \frac{q}{\eta}\langle I\rangle \tag{9.68}$$

Using (9.67) we can write

$$N(0) \approx \frac{\langle I\rangle^2}{f_c} \tag{9.69}$$

Consequently,

$$\frac{N(0)}{N^{(S)}_I} \approx \frac{\eta}{q}\frac{\langle I\rangle}{f_c} \tag{9.70}$$

We see that for a given linewidth f_c, the ratio of field intensity noise to shot noise equivalent power densities at $f = 0$ is proportional to the average field intensity $\langle I\rangle$. The crossover intensity I_0 for which both noise power densities are equal is estimated by

$$I_0 \approx \frac{q}{\eta}f_c \tag{9.71}$$

If the only two noise mechanisms present are shot noise and field-induced noise, then shot noise dominates for $\langle I\rangle < I_0$, and field-induced noise dominates for $\langle I\rangle > I_0$. It must be kept in mind, though, that while the power spectrum of shot noise is flat, field-induced noise is present only in a spectral band of width ~f_c centered at $f = 0$. All our estimates regarding the relative magnitude of the shot and field-induced noises are of course valid only near the origin of the frequency axis, as illustrated schematically in Figure 9.4. Note that we use f_c to denote both the field linewidth and the width of its intensity power spectrum, although in principle the latter is always wider by roughly a factor of $\sqrt{2}$.

Let us estimate I_0 for the common case of a light-emitting diode operating in the near IR and a silicon detector. In this case, our estimate of I_0 is quite realistic, since the statistics of a light-emitting diode is believed to be Gaussian. The linewidth f_c of such a source is typically on the order of 40 nm, which corresponds in the near IR to approximately 20 THz. For a silicon detector $\eta \sim 0.5$ A/W. Substituting these values in (9.71), we find, for this case,

$$I_0 \sim 6\ \mu\text{W} \tag{9.72}$$

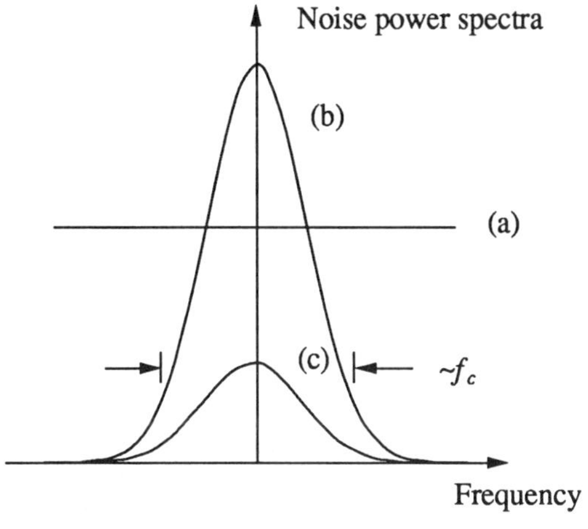

Figure 9.4 The schematic behavior of the power spectra of the shot and field induced noises: (a) The power spectrum of the shot noise; (b) the power spectrum of the field induced noise for $\langle I \rangle > I_0$; (c) the power spectrum of the field induced noise for $\langle I \rangle < I_0$.

This estimation shows that in systems using light-emitting diodes, the field-induced noise should be included in the noise budget whenever the power incident on the detector exceeds ~1 μW. If we use the same estimation for a laser diode with a linewidth of 1 GHz, we obtain

$$I_0 \sim 0.3\ \text{nW}$$

In other words, in coherent sources the field intensity noise overcomes the shot noise equivalent intensity noise at much lower thresholds. However, its spectral extension is significantly more limited.

Another two quantities that are worthwhile to compare to each other are the field and shot noise equivalent intensity noise variances for a detector combined with a low-pass electrical filter with bandwidth B. Since the shot noise equivalent intensity noise spectrum is flat, we obtain

$$\mathrm{Var}[\mathrm{I}^{(S)}] = \frac{q}{\eta} \langle \mathrm{I} \rangle B \tag{9.73}$$

On the other hand, for the field intensity variance we have

$$\mathrm{Var}[\mathrm{I}] = \begin{cases} \sim 2N(0)B & B \ll f_c, \\ \sim \langle \mathrm{I} \rangle^2 & B \gg f_c. \end{cases} \tag{9.74}$$

This state of affairs is illustrated in Figure 9.5.

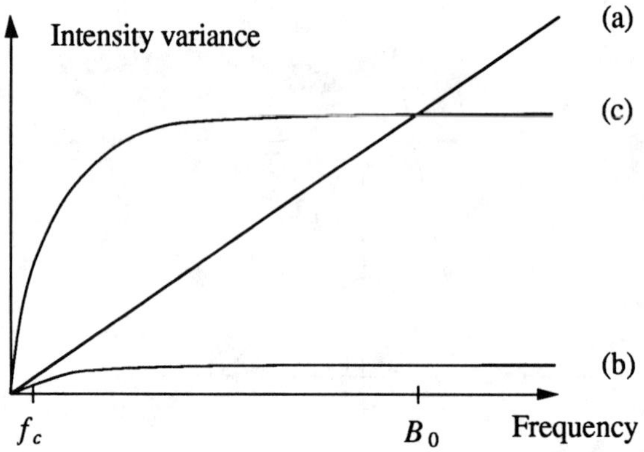

Figure 9.5 The variances of the field and the shot noise equivalent intensities: (a) the variance of the shot noise equivalent intensity $\mathrm{Var}[\mathrm{I}^{(S)}]$; (b) the variance of the optical intensity Var[I] for $\langle \mathrm{I} \rangle < I_0$; (c) the variance of the optical intensity Var[I] for $\langle \mathrm{I} \rangle > I_0$.

For $\langle \mathrm{I} \rangle > I_0$, there exists a bandwidth B_0 for which the two variances are equal. If $\langle \mathrm{I} \rangle \gg I_0$ (as in the case shown in Figure 9.4(c)), then B_0 is given by

$$B_0 = \frac{\eta}{2q} \langle \mathrm{I} \rangle = f_c \frac{\langle \mathrm{I} \rangle}{I_0} \tag{9.75}$$

For a laser diode with $f_c \sim 1$ GHz and $I_0 \sim 0.3$ nW, we get $B_0 \sim 10$ GHz for $<I> \sim 3$ nW.

9.7 SUMMARY

The fourth-order field correlation functions are necessary for the computation of the intensity noise power spectrum. Little can be said on correlation functions in general; however, here we are interested in correlation functions of stationary processes with a finite coherence time. The fourth-order correlation functions of such processes possess certain general asymptotic properties, which determine the analytic structure of their frequency-domain representation.

Two types of fields are of particular interest: Gaussian fields and random-phase fields. In the case of Gaussian fields, the fourth-order correlation functions can be expressed in terms of the second-order correlation functions. For the random-phase field, we express the fourth-order correlation functions in terms of the phase structure function. In both cases, the fourth-order correlation functions are expressed in terms of second-order quantities.

The field intensity power spectrum consists in general of a δ function which is superimposed on a smooth background of a finite width f_c. The magnitude of this singularity equals the average field intensity, and the smooth background is referred to as the intensity noise power spectrum. The area between the frequency axis and the intensity noise curve gives the intensity variance. The intensity power spectrum can be represented in terms of time-domain or in terms of frequency-domain quantities. For the evaluation of the intensity power spectrum, both representations seem to be equally useful. However, for the evaluation of the effect of a linear and time-independent network on the intensity power spectrum, the frequency-domain representation is advantageous.

In the discussion of the second-order field statistics we have argued that the complex representation of the Jones vector is more convenient for the analytic treatment. The same is true for the treatment of the fourth-order statistics. There is, however, a qualitative difference between the intensity power spectra of the real field and its associated complex field. The intensity power spectrum of the real field is composed of three components: a central line and two sidelobes, while the intensity power spectrum of the complex field contains only the central line. This difference is not important in practice, since the sidelobes of the real process are located at optical frequencies, and are filtered out by any known optical detector. Therefore, we are allowed to use the associated complex Jones vector for the computation of the intensity power spectrum.

In general, the field intensity noise is only one of the noise mechanisms that contribute to the system noise budget. It is of particular interest to compare the magnitude of the intensity noise to the magnitude of the shot noise, which is fundamental to the optical detection process. The field intensity noise power density is proportional to the second power of the intensity, while the shot noise power density is proportional simply to the intensity itself. Therefore, as the average intensity increases, the relative contribution of the intensity noise with respect to the shot noise increases. At a certain intensity, both power densities become equal. This intensity is on the order of 10 μW in the case of a typical light-emitting diode, and on the order of 0.1 nW for a single-mode laser diode. It must be remembered, though, that while the shot noise power spectrum is flat, the intensity noise power spectrum is limited to a spectral band of width f_c around the frequency axis origin.

References

[1] Goodman, J.W., *Statistical Optics*, New York: John Wiley & Sons, 1985.

[2] Weissman, Y., "The Theory of Source-Induced Noise in Optical Networks," *J. Opt. Soc. Am.*, Vol. 7B, No. 9, September 1990, pp. 1791-1800.

[3] Picinbono, B. and E. Boileau, "Higher-Order Coherence Functions of Optical Fields and Phase Fluctuations," *J. Opt. Soc. Am.*, Vol. 58, No. 6, June 1968, pp. 784-789.

[4] Pratt, W.K., *Laser Communication Systems*, New York: John Wiley & Sons, 1969.

Chapter 10
The Output Intensity Power Spectrum of Time-Independent Networks

In the preceding chapter we derived several expressions for the field and the intensity power spectra of given fields. In this chapter we will consider the problem of how to determine these quantities for a given source coupled to a linear and time-independent network, as shown in Figure 10.1.

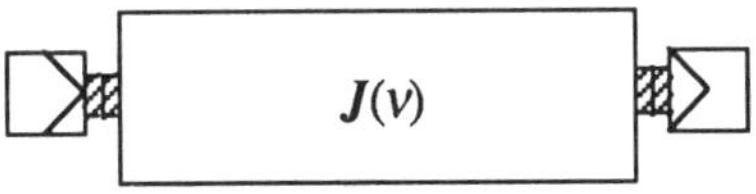

Figure 10.1 A generic optical system consisting of a source, a linear network, and a detector.

The linear network affects both the power and the intensity spectra of the source field. In particular, an intensity noise may be generated at the network output, even if the source is a pure phase-noise source, thus completely lacking intensity noise. We will assume that the second- and fourth-order statistics of the source field are known, as well as the network Jones (transfer) matrix $J(\nu)$. Given these quantities, we will attempt to compute the power and intensity spectra of the output optical field incident on the detector.

In general, the output intensity power spectrum depends on the source field correlation functions and the network Jones matrix in a complex and nonseparable way. This means that, in general, a different problem has to solved for each network-source combination. The general expressions can be simplified only for special cases, or in certain circumstances that allow us to use appropriate approximations. A factorization between the field and the network-dependent terms is possible in the incoherent limit. This is the case in which the fields that are superimposed at the various network nodes are statistically independent. It is possible to define two matrices that are characteristic of the network, so that in the incoherent limit the output intensity power spectrum is given by a plain formula involving only simple products between these matrices and the source-dependent terms. Consequently, once the network characteristic matrices are evaluated, the computation of the output intensity power spectrum for any given source becomes straightforward. Another interesting and important case is the problem of a random-phase source coupled to a dispersive waveguide. In this case it is possible to derive a simple analytic expression for the output intensity power spectrum.

This chapter illustrates the utility of the concepts we developed in the first part of the book. Once the network Jones matrix is determined, we can solve the signal analysis problem without any reference to the network structure, or to the structure of its components. All the relevant information about the network is contained in the four complex functions making up $\boldsymbol{J}(\nu)$.

10.1 THE POWER SPECTRUM OF THE OUTPUT FIELD

Let us consider a stationary optical source with a given correlation matrix $\boldsymbol{R}(\nu)$ coupled to the input of an optical network with a Jones matrix $\boldsymbol{J}(\nu)$, and compute the output field power spectrum $P'(\nu)$. We begin by deriving the relation between the input and output coherency matrices. The coherency matrix $\boldsymbol{G}'(\nu_1,\nu_2)$ of the output field is given by (8.28):

$$\boldsymbol{G}'(\nu_1,\nu_2) = \langle \boldsymbol{A}'(\nu_1)\boldsymbol{A}'^{\dagger}(\nu_2)\rangle$$

where $\boldsymbol{A}'(\nu)$ is the output Jones vector. By (2.15) we obtain

$$\begin{aligned}\boldsymbol{G}'(\nu_1,\nu_2) &= \boldsymbol{J}(\nu_1)\boldsymbol{G}(\nu_1,\nu_2)\boldsymbol{J}^{\dagger}(\nu_2)\\ &= \boldsymbol{J}(\nu_1)\boldsymbol{R}(\nu_1)\boldsymbol{J}^{\dagger}(\nu_1)\delta(\nu_1-\nu_2)\end{aligned} \tag{10.1}$$

Comparing (10.1) and (8.29), we conclude that

$$\boldsymbol{R}'(\nu) = \boldsymbol{J}(\nu)\boldsymbol{R}(\nu)\boldsymbol{J}^{\dagger}(\nu) \tag{10.2}$$

This is the relation we sought. Turning now to the field power spectrum, we obtain (8.35):

$$P'(\nu) = \mathrm{Tr}[\boldsymbol{J}(\nu)\boldsymbol{R}(\nu)\boldsymbol{J}^{\dagger}(\nu)] = \mathrm{Tr}[\boldsymbol{H}(\nu,\nu)\boldsymbol{R}(\nu)] \tag{10.3}$$

where we have introduced the H matrix,

$$H(\nu_1, \nu_2) = J^\dagger(\nu_1) J(\nu_2) \tag{10.4}$$

Equation (10.3) is the general expression for the output power spectrum of a stationary optical field coupled to a linear network characterized by a Jones matrix $J(\nu)$. We see that, in general, for the calculation of $P'(\nu)$ it is not enough to specify $P(\nu)$, since, for instance, $P'(\nu)$ may depend on the nondiagonal elements of $R(\nu)$, which do not contribute to $P(\nu)$.

For a separable source, the output power is given by

$$P'(\nu) = \sum_j T_j(\nu) P_j(\nu) \tag{10.5}$$

where we have denoted the diagonal elements of $H(\nu, \nu)$ with $T_j(\nu)$; i.e.,

$$T_j(\nu) = H_{jj}(\nu, \nu) \tag{10.6}$$

These functions may be called the *power transfer* functions for the corresponding polarization modes. In the case of a separable source, the output power spectrum is determined by the power spectra of the source in each of its polarization modes and the power transfer functions of the network.

10.2 THE POWER SPECTRUM OF THE OUTPUT FIELD INTENSITY

10.2.1 The General Expression

To compute the power spectrum of the output intensity, we apply (9.52) to the output field:

$$S'(f) = \int d^4\nu \langle \mathrm{Tr}[A'(\nu_1) A'^\dagger(\nu_2)] \mathrm{Tr}[A(\nu_3) A'^\dagger(\nu_4)] \rangle \delta(f - \nu_1 + \nu_2) \tag{10.7}$$

Using the relation (2.15) between the input and the output fields, we get

$$S'(f) = \int d^4\nu \langle \mathrm{Tr}[H(\nu_2, \nu_1) A(\nu_1) A^\dagger(\nu_2)] \mathrm{Tr}[H(\nu_3, \nu_4) A(\nu_3) A'^\dagger(\nu_4)] \rangle \delta(f - \nu_1 + \nu_2)$$

To proceed further, we have to express the right-hand side in terms of the input field correlation functions. This is achieved by replacing the trace operation with the explicit sums:

$$S'(f) = \sum_{n,m,k,j} \int d^4\nu H_{mn}(\nu_2, \nu_1) H_{jk}(\nu_4, \nu_3) G_{nmkj}(\nu_1, \nu_2, \nu_3, \nu_4) \tag{10.8}$$

We substitute now the analytic representation (9.20) for the fourth-order field correlation functions on the right-hand side, and obtain the three components $S'_1(f)$, $S'_2(f)$, and $S'_3(f)$ of the output intensity power spectrum $S'(f)$. The first component is given by

$$\begin{aligned} S'_1(f) &= \sum_{n,m,k,j} \int d^4\nu H_{mn}(\nu_2, \nu_1) H_{jk}(\nu_4, \nu_3) \Delta R_{nmkj}(\nu_1, \nu_1 - \nu_2, \nu_4) \\ &\cdot \delta(\nu_1 - \nu_2 + \nu_3 - \nu_4)\delta(f - \nu_1 + \nu_2) \end{aligned} \tag{10.9}$$

Performing the integration over ν_1 and ν_4 we get

$$S'_1(f) = \sum_{n,m,k,j} \int d^2\nu H_{mn}(\nu_1, \nu_1 + f) H^\dagger_{jk}(\nu_2, \nu_2 + f) \Delta R_{nmkj}(\nu_1 + f, f, \nu_2 + f) \tag{10.10}$$

where we have also renamed the integration variables. For the second term, we get

$$\begin{aligned} S'_2(f) &= \sum_{n,m,k,j} \int d^4\nu H_{mn}(\nu_2, \nu_1) H_{jk}(\nu_4, \nu_3) R_{nj}(\nu_1) R_{km}(\nu_3) \\ &\cdot \delta(\nu_1 - \nu_4)\delta(\nu_2 - \nu_3)\delta(f - \nu_1 + \nu_2) \end{aligned} \tag{10.11}$$

Performing the integrations over ν_1, ν_3 and ν_4, we obtain

$$S'_2(f) = \sum_{n,m,k,j} \int d\nu H_{mn}(\nu, \nu + f) H^\dagger_{jk}(\nu, \nu + f) R_{nj}(\nu + f) R_{km}(\nu) \tag{10.12}$$

This can be written compactly using matrix notation:

$$S'_2(f) = \int d\nu \mathrm{Tr}[\boldsymbol{H}^\dagger(\nu, \nu + f)\boldsymbol{R}(\nu)\boldsymbol{H}(\nu, \nu + f)\boldsymbol{R}(\nu + f)] \tag{10.13}$$

This term may be regarded as representing the Gaussian component of the output intensity power spectrum. It remains unaffected if the source is replaced by its Gaussian equivalent, and is always nonnegative. Finally, the third component is given by

$$S'_3(f) = \sum_{n,m,k,j} \int d^4\nu H_{mn}(\nu_2,\nu_1) H_{jk}(\nu_4,\nu_3) R_{nm}(\nu_1) R_{kj}(\nu_3) \delta(\nu_1 - \nu_2)\delta(\nu_3 - \nu_4) \quad (10.14)$$

This expression is readily transformed to

$$S'_3(f) = \left[\int d\nu \mathrm{Tr}[\boldsymbol{H}(\nu,\nu)\boldsymbol{R}(\nu)] \right]^2 \delta(f) \quad (10.15)$$

By (8.36) and (10.3), the term in the square brackets is simply the average <I'> of the output intensity.

We now write down the complete expression for the output intensity power spectrum [1]:

$$\begin{aligned} S'(f) &= \sum_{n,m,k,j} \int d^2\nu H_{mn}(\nu_1,\nu_1 + f) H^{\dagger}_{jk}(\nu_2,\nu_2 + f) \Delta R_{nmkj}(\nu_1 + f, f, \nu_2 + f) \\ &+ \int d\nu \mathrm{Tr}\{\boldsymbol{H}^{\dagger}(\nu,\nu + f)\boldsymbol{R}(\nu)\boldsymbol{H}(\nu,\nu + f)\boldsymbol{R}(\nu + f)\} + \langle I' \rangle^2 \delta(f) \end{aligned} \quad (10.16)$$

It is also useful to introduce the output intensity noise power spectrum $N'(f)$:

$$\begin{aligned} N'(f) &= S'_1(f) + S'_2(f) \\ &= \sum_{n,m,k,j} \int d^2\nu H_{mn}(\nu_1,\nu_1 + f) H^{\dagger}_{jk}(\nu_2,\nu_2 + f) \Delta R_{nmkj}(\nu_1 + f, f, \nu_2 + f) \\ &+ \int d\nu \mathrm{Tr}\{\boldsymbol{H}^{\dagger}(\nu,\nu + f)\boldsymbol{R}(\nu)\boldsymbol{H}(\nu,\nu + f)\boldsymbol{R}(\nu + f)\} \end{aligned} \quad (10.17)$$

10.2.2 Simplifications of the General Case

The general formula (10.16) is rather complicated. In practice, however, we seldom deal with the general case, and in many special cases of practical importance, (10.16) is considerably simplified. Perhaps the most obvious simplification occurs if the source is Gaussian, since then $\Delta R_{nmkj}(\nu_1, f, \nu_4)$ vanishes, and we are left with

$$N'(f) = \int d\nu \mathrm{Tr}[\boldsymbol{H}^{\dagger}(\nu,\nu + f)\mathbf{R}(\nu)\boldsymbol{H}(\nu,\nu + f)\boldsymbol{R}(\nu + f)] \quad (10.18)$$

Another simplification occurs when the source is separable. In this case, we have

$$N'(f) = \sum_n \int d^2\nu H_{nn}(\nu_1,\nu_1 + f)H^\dagger{}_{nn}(\nu_2,\nu_2 + f)\Delta R_{nnnn}(\nu_1 + f,f,\nu_2 + f)$$
$$+ \sum_{n,m} \int d\nu\, |H_{nm}(\nu,\nu + f)|^2 P_n(\nu)P_m(\nu + f) \tag{10.19}$$

It follows, therefore, that for a separable Gaussian source (for which the first term vanishes), the power spectra (lineshapes) of each one of the polarization modes and the network Jones matrix determine the output intensity noise power spectrum. If the source is polarized, so that only the jth polarization mode is present in the source field, we obtain

$$N'(f) = \int d^2\nu H_{jj}(\nu_1,\nu_1 + f)H^*{}_{jj}(\nu_2,\nu_2 + f)\Delta R_{jjjj}(\nu_1 + f,f,\nu_2 + f)$$
$$+ \int d\nu\, |H_{jj}(\nu,\nu + f)|^2 P_j(\nu)P_j(\nu + f) \tag{10.20}$$

Finally, let us consider the case of the polarization-degenerate network, for which

$$\boldsymbol{H}(\nu_1,\nu_2) = H(\nu_1,\nu_2)\boldsymbol{I}$$

We will refer to $H(\nu_1,\nu_2)$ as the *H function* of the network. Using this relation in (10.17), we obtain

$$N'(f) = \sum_{n,k} \int d^2\nu H(\nu_1,\nu_1 + f)H^*(\nu_2,\nu_2 + f)\Delta R_{nnkk}(\nu_1 + f,f,\nu_2 + f)$$
$$+ \int d\nu\, |H(\nu,\nu + f)|^2 \mathrm{Tr}[\boldsymbol{R}(\nu)\boldsymbol{R}(\nu + f)] \tag{10.21}$$

For a separable source coupled to a degenerate network, we can write

$$N'(f) = N'_1(f) + N'_2(f) \tag{10.22}$$

where $N'_j(f)$ is the output intensity noise power spectrum associated with the jth polarization mode:

$$N'_j(f) = \int d^2\nu H(\nu_1,\nu_1 + f)H^*(\nu_2,\nu_2 + f)\Delta R_{jjjj}(\nu_1 + f,f,\nu_2 + f)$$
$$+ \int d\nu\, |H(\nu,\nu + f)|^2 P_j(\nu)P_j(\nu + f) \tag{10.23}$$

In this case, it is therefore possible to consider each polarization mode separately and to determine the output intensity noise spectrum by simply superimposing the results.

The analysis can be often simplified by the observation that multiplicative exponential factors in the network Jones matrix do not affect the output intensity noise power spectrum, and therefore can be disregarded. Suppose that

$$J(\nu) = e^{i\alpha\nu} J'(\nu)$$

where α is a frequency-independent constant. Then

$$\mathbf{H}(\nu, \nu + f) = e^{-i\alpha\nu} J'^{\dagger}(\nu) e^{i\alpha(\nu+f)} J'(\nu + f) = e^{i\alpha f} H'(\nu, \nu + f)$$

where H' is the H matrix corresponding to J'. Similarly,

$$H^{\dagger}(\nu, \nu + f) = e^{-i\alpha f} H'^{\dagger}(\nu, \nu + f)$$

and obviously

$$H(\nu, \nu) = H'(\nu, \nu)$$

Inspection of (10.17) reveals that the exponential term $e^{i\alpha\nu}$ cancels out in all three terms containing $S'_{\cdot}(f)$.

The cancellation of multiplicative factors of the form $e^{i\alpha\nu}$ in the expression for the output intensity power spectrum is intuitively obvious. When transformed into the time domain, such factors represent a constant time delay between the source and the network input, or between the network output and the detector. Such time delays cannot affect the signal statistics.

10.3 A PHASE-NOISE SOURCE COUPLED TO A DISPERSIVE WAVEGUIDE

The field that appears at the output of a linear network coupled to a random-phase source will *not* generally be a random-phase field. Therefore, a nonvanishing intensity noise may be expected in the output field, even if the intensity noise of the input field vanishes. The generation of intensity noise by the linear filtering of a random-phase field has been called in the literature "phase-to-intensity noise conversion."

One of the striking examples of phase to intensity noise conversion is the generation of intensity noise by a random-phase source coupled to a dispersive waveguide [2]. This phenomenon is of considerable importance in long-distance communication lines with nonregenerative repeaters, such as optical amplifiers. Theory predicts that the field-induced noise becomes dominant in gigabit-per-second links with source linewidths larger than several tens of megahertz and total line dispersion exceeding several thousand picoseconds per nanometer [2]. The intrinsic dispersion of fibers, which will be denoted by m_d, in the 1.5μ band is on the order of 20 ps/(nm · km). Therefore, a 500-km communication link will possess an accumulated dispersion of 10^4 ps/nm.

In this section we will apply the expressions derived for the output intensity noise power spectrum to the problem of the phase-noise source coupled to a lossless but dispersive waveguide. For simplicity, we will assume that the waveguide is polarization-degenerate, and that the source is separable.

10.3.1 A Degenerate Dispersive Waveguide and Its *H* Function

Let us write the Jones matrix of the waveguide in the form

$$J(\nu) = e^{-i\beta L}\boldsymbol{I}$$

where β is the propagation "constant":

$$\beta = \frac{2\pi n \nu}{c} \tag{10.24}$$

In general, the (effective) refraction index n depends on the frequency ν. This dependence is due to the waveguide material properties as well as to its guiding characteristics. We assume here that in the spectral region of the source, β can be approximated by its second-order Taylor expansion:

$$\beta \approx \beta_0 + \beta_1 \nu + \beta_2 \nu^2 \tag{10.25}$$

Within this approximation, the H function is given by

$$\begin{aligned} H(\nu,\nu+f) &= \exp\{i(\beta_0 + \beta_1\nu + \beta_2\nu^2)L - i[\beta_0 + \beta_1(\nu+f) + \beta_2(\nu+f)^2]\} \\ &= \exp[-i\beta_1 fL - i\beta_2 fL(2\nu + f)] \end{aligned} \tag{10.26}$$

For the computation of the first term in (10.20), we need the product of two H functions:

$$H(\nu_1,\nu_1+f)H^*(\nu_2,\nu_2+f) = \exp[2\pi i(\nu_1 - \nu_2)\tau] \tag{10.27}$$

where we have introduced

$$\tau = \frac{1}{\pi} \beta_2 f L \tag{10.28}$$

The intrinsic dispersion of the fiber m_d is related to β_2 by [2]

$$\beta_2 = \frac{\pi\lambda^2}{c} m_d \tag{10.29}$$

We can express τ in terms of m_d as follows:

$$\tau = \frac{\lambda^2 m_d f L}{c} \tag{10.30}$$

10.3.2 Computation of the First Term in the Output Noise Power Spectrum

Since we deal here with a separable field coupled to a degenerate network, we may use (10.22) and treat each one of the polarization modes separately. For simplicity, we drop the polarization mode index. To evaluate $S'_1(f)$, we use (10.9) and (10.27):

$$\begin{aligned} S'_1(f) &= \int d^2\nu \exp[2\pi i(\nu_1 - \nu_2)\tau] \Delta R(\nu_1 + f, f, \nu_2 + f) \\ &= \int d^2\nu \exp[2\pi i(\nu_1 - \nu_2)\tau] \Delta R(\nu_1, f, \nu_2) \end{aligned} \tag{10.31}$$

The last integral is essentially a double Fourier integral, and can be expressed in terms of the time-domain representation of the auxiliary correlation function:

$$S'_1(f) = \int dt\ e^{2\pi i f t} \Delta R(-\tau, t, \tau) \tag{10.32}$$

Thus, the evaluation of $S'_1(f)$ is reduced to the computation of a Fourier transform. The function in the integrand is derived from (9.35):

$$\Delta R(-\tau,t,\tau) = \langle I\rangle^2 \exp[-\Lambda(\tau) - \Lambda(t) + \tfrac{1}{2}\Lambda(t-\tau) + \tfrac{1}{2}\Lambda(t+\tau)] - e^{-\Lambda(t)} - e^{-\Lambda(\tau)} \tag{10.33}$$

In order to proceed, we assume that $\Lambda(t) = \gamma|t|$ (Eq. (8.71)). Since $S'_1(f)$ is an even function of f, it is sufficient to compute it for $f \geq 0$. Therefore in what follows we assume that $\tau \geq 0$. With these assumptions we get

$$\Delta R(-\tau,t,\tau) = \begin{cases} -\langle I\rangle^2 e^{-\gamma t} & \text{for } |t| \geq \tau, \\ -\langle I\rangle^2 e^{-\gamma\tau} & \text{for } |t| \leq \tau. \end{cases} \qquad (10.34)$$

This function is displayed schematically in Figure 10.2.

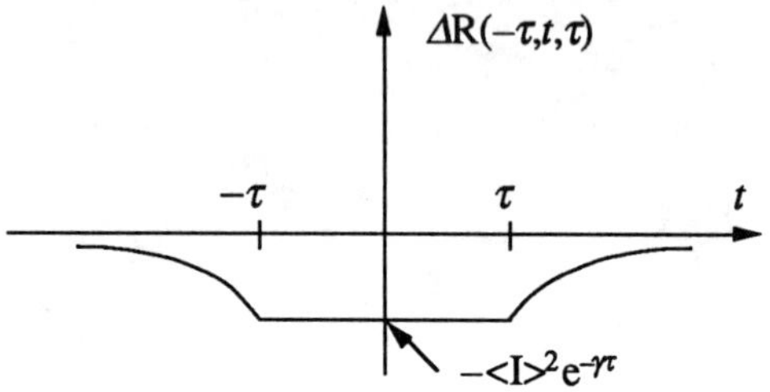

Figure 10.2 Graphical representation of (10.34).

Accordingly, the integral (10.32) is a sum of three terms:

$$\langle I\rangle^{-2} S'_1(f) = -e^{-\gamma\tau}\int_{-\tau}^{-\tau} dt\, e^{2\pi i f t} + \int_{-\infty}^{-\tau} dt\, e^{\gamma t + 2\pi i f t} + \int_{\tau}^{\infty} dt\, e^{-\gamma t + 2\pi i f t}$$

The evaluation of the integrals on the right-hand side is straightforward, and a few additional algebraic manipulations yield

$$S'_1(f) = -\frac{\langle I\rangle^2 \gamma e^{-\gamma|\tau|}}{4\pi^2 f^2 + \gamma^2}\left[\frac{\gamma}{\pi f}\sin(2\pi f|\tau|) + 2\cos(2\pi f|\tau|)\right] \qquad (10.35)$$

We have taken everywhere the absolute value of τ, since $S'_1(f)$ must be an even function of f.

10.3.3 The Output Intensity Noise Power Spectrum

Since, in the present case, $|H(\nu + f, \nu)|^2 = 1$, $S'_2(f)$ is simply

$$S'_2(f) = \int d\nu R(\nu)R(\nu + f) = \langle I\rangle^2 \int dt\, e^{2\pi i f t - \gamma|t|} = \frac{2\langle I\rangle^2 \gamma}{4\pi^2 f^2 + \gamma^2} \qquad (10.36)$$

The output intensity noise power spectrum $N'(f)$ is therefore given by

$$N'(f) = S'_1(f) + S'_2(f)$$

$$= \frac{2\langle I\rangle^2\gamma}{4\pi^2 f^2 + \gamma^2}\left[1 - \frac{\gamma e^{-\gamma|\tau|}}{2\pi f}\sin(2\pi f|\tau|) - e^{-\gamma|\tau|}\cos(2\pi f|\tau|)\right] \tag{10.37}$$

We introduce the dimensionless parameters

$$\alpha = \frac{\beta_2\gamma^2 L}{\pi}, \qquad x = \frac{f}{\gamma} \tag{10.38}$$

These parameters can be used to rewrite (10.37) in a dimensionless form:

$$\gamma RIN'(x) = \frac{2}{1 + (2\pi x)^2}\left\{1 - e^{-\alpha x}\left[\frac{\sin(2\pi\alpha x^2)}{2\pi|x|} + \cos(2\pi\alpha x^2)\right]\right\} \tag{10.39}$$

where $RIN'(x)$ is the relative noise power spectrum of the output intensity, as defined in (9.65). Let us express α in terms of the total dispersion $m_d L$ and γ:

$$\alpha = \frac{\lambda^2\gamma^2 m_d L}{c} \tag{10.40}$$

If we measure λ in microns, $m_d L$ in picoseconds per nanometer and γ in gigahertz, then α is given by the formula

$$\alpha = 3.33\cdot 10^{-6}\lambda^2\gamma^2 m_d L \tag{10.41}$$

Two experimental curves [2] of the output intensity noise power spectrum are shown in Figure 10.3. The spectra were taken with a DFB laser emitting at $\lambda = 1.536\mu$ with a linewidth of ~300 MHz. In order to compare our theoretical expression with these results, we have plotted $\gamma RIN'(x)$ for three values of α in Figure 10.4.

According to (8.72), the linewidth at half height is equal to $\gamma/2\pi$, so that the value of γ corresponding to a linewidth of ~300 MHz is ~2 GHz. Taking λ = 1.536μ, the following values of α are obtained from (10.40):

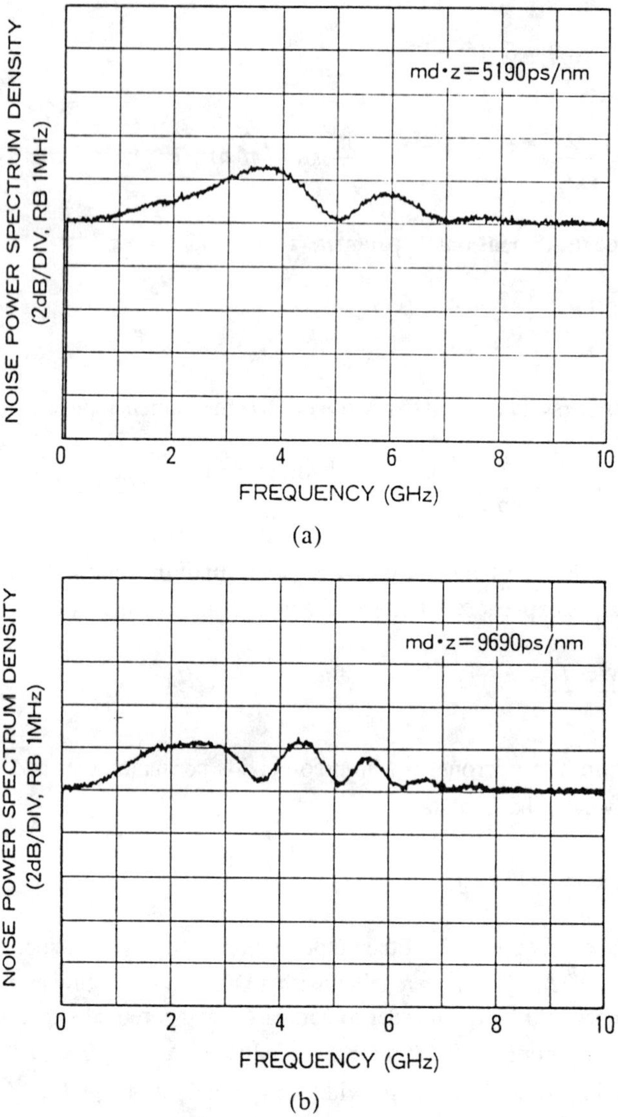

Figure 10.3 Experimental curves showing the output noise power spectrum of a phase-noise source coupled to a dispersive fiber [2] (© IEEE, 1990).

Table 10.1 Values of the parameter α and the crossover intensity for three values of the total dispersion.

$m_d L$	α	I_0 (nW)
$5 \cdot 10^3$	0.1573	3.2
10^4	0.3146	1.6
$2 \cdot 10^4$	0.6292	0.87

Comparing these values of α to the values chosen for curves (a) and (b) in Figure 10.4, we conclude that these curves correspond to the same experimental parameters as the curves shown in Figure 10.3. Even though a quantitative comparison between the theoretical and the experimental curves cannot be made, there is a qualitative agreement in the positions of their corresponding extrema (in comparing Figures 10.3 and 10.4, it must be remembered that each division in Figure 10.4 corresponds to 2 GHz). A different, approximate expression for the output intensity noise power spectrum which is also in qualitative agreement with experiment has been derived by Yamamoto et al. [2].

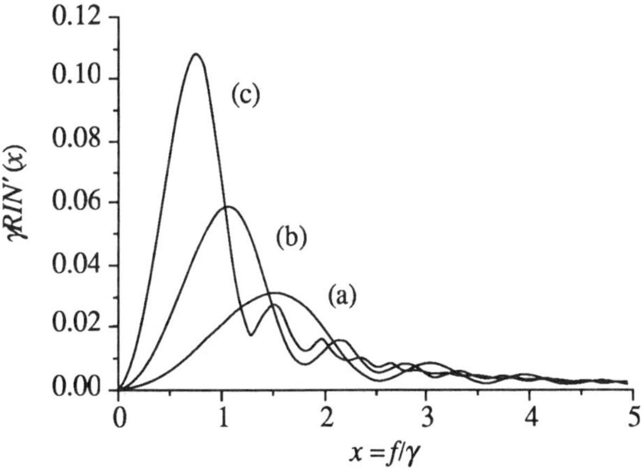

Figure 10.4 A plot of the output relative intensity noise power spectrum (Eq. (10.39)) for three values of α: (a) $\alpha = 0.1573$ (corresponding to $m_d L = 5{,}000$ ps/nm); (b) $\alpha = 0.3146$ (corresponding to $m_d L = 10{,}000$ ps/nm); (c) $\alpha = 0.6292$ (corresponding to $m_d L = 20{,}000$ ps/nm).

It is evident from Figure 10.4 that the maximum of $N(f)$ increases sharply with α. The value of the crossover intensity I_0 at which the maximum N_m of $N(f)$ reaches the value of the equivalent shot noise is determined by

$$\frac{N_m I_0^2}{\gamma} = \frac{q I_0}{\eta}$$

Using the common value of 0.5 A/W for η, we have included in the third column of Table 10.1 the corresponding values of I_0. For intensities larger than I_0, the maximum of $N(f)$ will exceed the level of the equivalent shot noise power density.

10.4 FREQUENCY-PERIODIC NETWORKS AND THE INCOHERENT LIMIT

10.4.1 Discrete and Frequency-Periodic Networks

A very useful and simple model of optical networks is the *discrete network* model. This model may be used whenever the guided field coherence length $L_c = c\tau_c$ is much larger than the typical size L_s of the network components, but smaller than the length of the network waveguide interconnects, as shown schematically in Figure 10.5. Qualitatively speaking, the scattering parameters of a component with size L_s will change significantly only if the frequency is changed by a quantity on the order of c/L_s. On the other hand, the width of the power spectrum of the field is on the order of f_c. Therefore, it may be expected that if $L_c \gg L_s$, then the frequency variation of the scattering parameters within the spectral range of the source would be negligible. Since the spectral range of the source is the only frequency range that is relevant to our analysis, if $L_c \gg L_s$ we may neglect completely the frequency dependence of the scattering parameters, and treat them as if they are frequency-independent. If, in addition, we assume that the network waveguides are ideal (no reflections and no polarization modes coupling), we obtain the discrete network model.

The discrete network transfer matrix $\boldsymbol{J}(\nu)$ is in the form

$$\boldsymbol{J}(\nu) = \boldsymbol{F}[\exp(2\pi i\nu\tau_1), \exp(2\pi i\nu\tau_2), \ldots, \exp(2\pi i\nu\tau_N)] \tag{10.42}$$

where the quantities τ_k are the various delay times associated with the waveguides, and $\boldsymbol{F}$ does not have any frequency dependence in addition to what is shown explicitly in (10.42). (If a waveguide is not degenerate, there are two distinct delay times associated with it, each representing the delay time of the corresponding polarization mode.) All four networks that we analyzed in Chapter 5 are discrete, and therefore had transmission matrices in the form of (10.42).

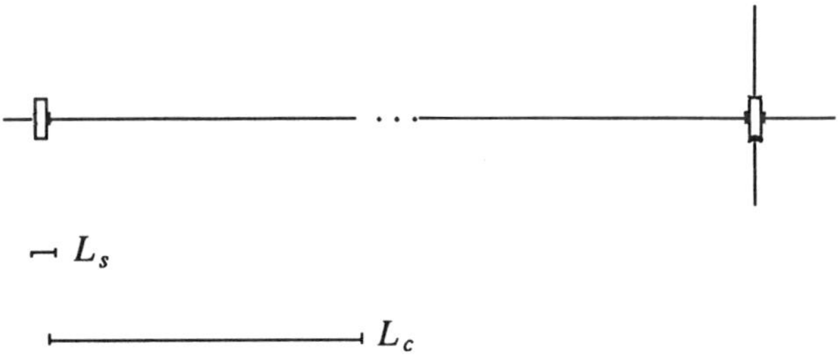

Figure 10.5 A segment of a network for which the discrete network model may be applied.

In practice, Jones matrices in the form of (10.42) may be regarded as having a periodic frequency dependence. Since the delay times $\tau_1, \tau_2, ..., \tau_N$ are always specified with a finite precision, we can assume without loss of generality that they are rationally dependent; i.e., there is a set of integers $n_1, n_2, ..., n_N$ and a time T such that

$$\tau_k = n_k T \tag{10.43}$$

We can thus write $\boldsymbol{J}(\nu)$ in the form

$$\boldsymbol{J}(\nu) = \boldsymbol{F}[\exp(2\pi i \nu n_1 T), \exp(2\pi i \nu n_2 T), ..., \exp(2\pi i \nu n_N T)] \tag{10.44}$$

from which it is evident that its frequency dependence is periodic with a period $\Delta f = 1/T$. Obviously, (10.43) is not a unique definition of T, since if T satisfies (10.43), then for any integer m, T/m will also satisfy (10.43) with the set of integers $mn_1, mn_2, ..., mn_N$. We will refer to the largest T that satisfies (10.43) as the *common delay* of the network. The time delays between the fields that are superimposed at the various nodes of a discrete network are whole multiples of its common delay time T.

In discrete networks, especially in fiber-optic networks, it often happens that the coherence time τ_c of the source is much shorter than the common delay time T. When this condition is fulfilled, we say that the network operates in the *incoherent limit*, or is simply *incoherent*. The reason for this name is that in such a network the fields that are superimposed at the various network nodes are delayed, one with respect to the other, by time periods that are much longer then τ_c. Such fields are statistically independent, or in the optical nomenclature, incoherent. We would like to emphasize that an incoherent network is, by our definition, also discrete, and therefore frequency-periodic.

The evaluation of $N'(f)$ for incoherent networks is greatly facilitated. To derive the appropriate approximations for this case, we will need a certain mathematical tool, which we will call the *decomposition theorem* [1].

10.4.2 The Decomposition Theorem

Suppose we have two functions, $w(\nu)$ and $F(\nu)$, with the following properties: 1. $w(\nu)$ is periodic in ν with a period Δf, and 2. the value of $F(\nu)$ does not change significantly when the argument ν is changed by an amount smaller then Δf; i.e.,

$$F(\nu) \approx F(\nu + \varepsilon) \tag{10.45}$$

for $|\varepsilon| < \Delta f$, or, in other words, $F(\nu)$ is smooth on the scale of Δf. If these conditions are fulfilled, then

$$\int d\nu w(\nu) F(\nu) \approx \{w\}_\nu \int d\nu F(\nu) \tag{10.46}$$

where

$$\{w\}_\nu = \frac{1}{\Delta f} \int_{[\Delta f]} d\nu w(\nu) \tag{10.47}$$

The integration in (10.47) is carried over the period Δf of $w(\nu)$. The quantity $\{w\}_\nu$ can be considered as the average of $w(\nu)$.

Proof. Let us express the integral on the left-hand side of (10.46) as a sum of an infinite number of integrals, each extending over an interval of length Δf:

$$\int d\nu w(\nu) F(\nu) = \sum_k \int_{k\Delta f}^{k\Delta f + \Delta f} d\nu w(\nu) F(\nu)$$

Using property 2, we can approximate this as

$$\int d\nu w(\nu) F(\nu) \approx \sum_k F(k\Delta f) \int_{k\Delta f}^{k\Delta f + \Delta f} d\nu w(\nu) = \{w\}_\nu \Delta f \sum_k F(k\Delta f)$$

In view of (10.45), we may approximate the sum over k by an integral, obtaining finally

$$\int d\nu w(\nu)F(\nu) \approx \{w\}_\nu \int d\nu F(\nu)$$

which is what we wanted to prove.

The decomposition theorem can be easily generalized for functions with any number of arguments. Suppose that we have two n-variable functions $w(\nu_1,\nu_2,...,\nu_n)$ and $F(\nu_1,\nu_2,...,\nu_n)$, such that (1) $w(\nu_1,\nu_2,...,\nu_n)$ is periodic in all its arguments with a common period Δf, i.e.

$$w(\nu_1,\nu_2,...,\nu_k,...,\nu_n) = w(\nu_1,\nu_2,...,\nu_k + \Delta f,...,\nu_n)$$

for any k , and (2) The value of $F(\nu_1,\nu_2,...,\nu_n)$ does not change significantly when any of its arguments is changed by an amount much smaller in absolute value than Δf. Under these circumstances, we can prove that

$$\int d^n\nu w(\nu_1,\nu_2,...,\nu_n)F(\nu_1,\nu_2,...,\nu_n) \approx \{w\}_\nu \int d^n\nu F(\nu_1,\nu_2,...,\nu_n) \tag{10.48}$$

where

$$\{w\}_\nu = \frac{1}{\Delta f^n} \int_{[\Delta f]^n} d^n\nu w(\nu_1,\nu_2,\ldots,\nu_n) \tag{10.49}$$

and $[\Delta f]^n$ represents an n-dimensional cube. The proof is a straightforward generalization of the proof that we have presented above for the single-variable case.

10.4.3 The Output Average Intensity in the Incoherent Limit Approximation

To get acquainted with the use of the decomposition theorem, let us compute the output average intensity <I'> in the incoherent limit. To accomplish this, we apply the decomposition theorem to (10.3). The role of the periodic function is played by $\boldsymbol{H}(\nu,\nu)$, and that of the smooth function by $\boldsymbol{R}(\nu)$ (see below). Thus, in the incoherent limit we obtain

$$\langle I' \rangle = \mathrm{Tr}[\{\boldsymbol{H}(\nu,\nu)\}_\nu \mathbf{R}(0)]$$

The matrix **R**(0) is hermitian, and can always be diagonalized by choosing an appropriate polarization mode basis. Assuming that this basis is used, we can cast $\langle I' \rangle$ in the form

$$\langle I' \rangle = \sum_j \{T_j(\nu)\}_\nu \langle I_j \rangle \tag{10.50}$$

where $\langle I_j \rangle$ is the average source intensity in the jth polarization mode, and $T_j(\nu)$ are the network power transfer functions (10.6). There is a striking similarity between (10.5) and (10.50); however, (10.50) was derived without assuming that the source is separable. In view of (10.50), it is possible in the incoherent limit to define average power transmissions $\{T_j(\nu)\}_\nu$ for the network, which can be used for the computation of the output average intensity with any given source.

10.4.4 The Power Spectrum of the Optical Intensity Noise in the Incoherent Limit Approximation

The approximation of the output intensity power spectrum in the incoherent limit is derived by applying the decomposition theorem to (10.17). Since an incoherent network is frequency-periodic, the various elements of the H matrix are frequency-periodic as well, and the first condition of the decomposition theorem is fulfilled. Let us consider now the second condition.

In (10.17), there are two functions that must be smooth on the scale of the frequency period Δf: the elements of $\mathbf{R}(\nu)$ and the auxiliary correlation functions $\Delta R_{nmkj}(\nu_1,\nu_2,\nu_3)$. Since the time-domain counterparts of these functions vanish as the absolute value of their arguments becomes much larger than τ_c, we may expect that the values of these functions vary significantly only if their arguments are varied by a quantity on the order of the field linewidth f_c. In the incoherent limit this quantity is much larger than Δf, and therefore both $R_{kj}(\nu)$ and $\Delta R_{nmkj}(\nu_1,\nu_2,\nu_3)$ may be expected to be smooth on the scale of Δf. These arguments are of course qualitative, but they are quite reliable in practice.

Application of the decomposition theorem to (10.17) yields

$$\begin{aligned} N'(f) = &\sum_{n,m,k,j} \{H_{mn}(\nu,\nu+f)\}_\nu \{H^\dagger_{jk}(\nu,\nu+f)\}_\nu \int d^2\nu \Delta R_{nmkj}(\nu_1,f,\nu_2) \\ &+ \sum_{n,m,k,j} \{H^\dagger_{mn}(\nu,\nu+f) H_{kj}(\nu,\nu+f)\}_\nu \int d\nu R_{nk}(\nu) R_{jm}(\nu+f) \end{aligned} \tag{10.51}$$

We would like to point out that the averaging in (10.51) is performed with respect to ν only, and the factors $\{H_{mn}(\nu,\nu+f)\}_\nu$ and $\{H^\dagger_{mn}(\nu,\nu+f)H_{kj}(\nu,\nu+f)\}_\nu$ are still periodic functions of f with a period Δf.

The general expression is considerably simplified for separable sources. Applying the decomposition theorem to (10.9), we get

$$N'(f) = \sum_{n} |\{H_{nn}(\nu,\nu+f)\}_\nu|^2 \int d^2\nu \Delta R_{nnnn}(\nu_1 + f, f, \nu_2 + f)$$
$$+ \sum_{n,m} \{|H_{nm}(\nu,\nu+f)|^2\}_\nu Q_{nm}(f) \qquad (10.52)$$

where we have introduced

$$Q_{nm}(f) = \int d\nu P_n(\nu) P_m(\nu+f) \qquad (10.53)$$

From (9.56) we derive the following expression for the power spectrum $N'(f)$ of the output intensity noise:

$$N'(f) = \sum_{n} |\{H_{nn}(\nu,\nu+f)\}_\nu|^2 N_n(f)$$
$$+ \sum_{n,m} [\{|H_{nm}(\nu,\nu+f)|^2\}_\nu - |\{H_{nm}(\nu,\nu+f)\}_\nu|^2 \delta_{nm}] Q_{nm}(f) \qquad (10.54)$$

where $N_n(f)$ is the source intensity noise power spectrum associated with its nth polarization mode. Equation (10.54) is very useful, since it relates the output intensity noise spectrum to the experimentally available functions $N_n(f)$ and $Q_{nm}(f)$. The source intensity noise power spectra $N_n(f)$ can be measured with a suitable polarizer and a radio-frequency spectrum analyzer, and the functions $Q_{nm}(f)$ can be computed from the lineshapes $P_n(\nu)$ with the help of (10.53). These are the only source characteristics that are necessary to compute $N'(f)$ in (10.54).

Due to the presence of the second term in (10.54), we see that, in general, the output intensity noise does not vanish even if the source intensity noise vanishes. This term is nonnegative, since by Schwartz's inequality,

$$\{|H_{nn}(\nu,\nu+f)|^2\}_\nu - |\{H_{nn}(\nu,\nu+f)\}_\nu|^2 \geq 0$$

A simple reasoninig leads to the conclusion that, from all sources with *given* lineshapes $P_j(\nu)$, those that behave as random-phase sources will always yield the smallest value of $N'(f)$. The reasonong runs as follows. For all sources with given lineshapes, the second term in (10.54) is the same, since the functions $Q_{nm}(f)$ are uniquely derived from the lineshapes $P_j(\nu)$. Therefore, the smallest value of $N'(f)$ will be attained by those sources for which the first term in (10.54) is smallest. The smallest possible value of this term is 0, and this value is actually attained by random-phase sources, for which $N_n(f)$ vanishes. This conclusion is valid for any network, but is limited to the incoherent limit situation.

In many cases, we deal with polarization-degenerate networks, and it is worthwhile to derive an explicit formula for the output intensity noise power spectrum for this case. Applying the decomposition theorem to (10.23) and using (10.22), we obtain

$$N'_n(f) = \sum_n N'_n(f) \tag{10.55}$$

where

$$N'_n(f) = |\{H(\nu,\nu+f)\}_\nu|^2 N_n(f) + [\{|H(\nu,\nu+f)|^2\}_\nu - |\{H(\nu,\nu+f)\}_\nu|^2] N^{(G)}{}_n(f) \tag{10.56}$$

In (10.56), $N_n(f)$ and $N^{(G)}{}_n(f)$ are the source and the equivalent Gaussian source noise power spectra, respectively, associated with the nth polarization mode. We conclude that in the case of a separable source coupled to a polarization-degenerate network, we can consider each of the polarization modes independently, and use the H function to represent the effect of the network. Thus, in this case, the noise evaluation problem is reduced to two independent scalar problems.

10.5 NETWORK PARAMETERS FOR THE CHARACTERIZATION OF THE OUTPUT INTENSITY NOISE IN THE INCOHERENT LIMIT

10.5.1 The Network Characteristic Matrices and Noise Factors

For a given frequency-periodic network, we can define the two following matrices:

$$K_{nm}(f) = |\{H_{nm}(\nu,\nu+f)\}_\nu|^2 \delta_{nm} \tag{10.57}$$

and

$$L_{nm}(f) = \{|H_{nm}(\nu,\nu+f)|^2\}_\nu \tag{10.58}$$

to which we will refer as the K and the L network *characteristic matrices*. The elements of both of these matrices are periodic functions in f with a period of Δf. In terms of these matrices, we can cast (10.54) in the form

$$N'(f) = \sum_{n} K_{nn}(f)N_n(f) + \sum_{n,m} [L_{nm}(f) - K_{nm}(f)]Q_{nm}(f) \tag{10.59}$$

In the case of the random-phase source, $N_n(f)$ vanishes, so that (10.59) is reduced to

$$N'(f) = \sum_{n,m} [L_{nm}(f) - K_{nm}(f)]Q_{nm}(f) \quad \text{(random-phase source)} \tag{10.60}$$

On the other hand, for a Gaussian source (Eq. (9.57)),

$$N_n(f) = Q_{nn}(f)$$

so that

$$N'(f) = \sum_{n,m} L_{nm}(f)Q_{nm}(f) \quad \text{(Gaussian source)} \tag{10.61}$$

For a polarization-degenerate network, we may define K and L *characteristic functions*:

$$K(f) = |\{H(\nu, \nu+f)\}_\nu|^2 \tag{10.62}$$

and

$$L(f) = \{|H(\nu, \nu+f)|^2\}_\nu \tag{10.63}$$

The polarization-degenerate version of (10.59) reads

$$N'(f) = K(f)N(f) + [L(f) - K(f)]N^{(G)}(f) \tag{10.64}$$

In network analysis, it is often preferable to deal with relative rather than absolute noises. The relative noise power spectrum $RIN'(f)$ of the output intensity is obtained by dividing (10.64) by $\langle I'\rangle^2$:

$$RIN'(f) = \frac{K(f)}{\{T(\nu)\}_\nu^2} RIN(f) + \frac{L(f) - K(f)}{\{T(\nu)\}_\nu^2} RIN^{(G)}(f) \tag{10.65}$$

where we have used (10.50). In practice, we most often deal with either Gaussian or random-phase sources. Consequently, it is convenient to define two network *noise factors*:

$$NNF^{(G)}(f) = \frac{L(f)}{\{T(\nu)\}_\nu^2} \tag{10.66}$$

and

$$NNF^{(rp)}(f) = \frac{L(f) - K(f)}{\{T(\nu)\}_\nu^2} \tag{10.67}$$

We can write (10.65) in terms of the network noise factors as follows:

$$RIN'(f) = [NNF^{(G)}(f) - NNF^{(rp)}(f)]RIN(f) + NNF^{(rp)}(f)RIN^{(G)}(f) \tag{10.68}$$

For the case of a random-phase source, $RIN(f) = 0$, so that

$$RIN'(f) = NNF^{(rp)}(f)RIN^{(G)}(f) \tag{10.69}$$

For the case of the Gaussian source, $RIN(f) = RIN^{(G)}(f)$, and we obtain

$$RIN'(f) = NNF^{(G)}(f)RIN^{(G)}(f) \tag{10.70}$$

Equations (10.69) and (10.70) are the polarization-degenerate versions of (10.60) and (10.61), respectively.

10.5.2 The Output Intensity Variance

For certain purposes, such as the computation of the bit-error probability in a digital communication system, it is necessary to know the variance of the detector current fluctuations. The field-induced component of the current variance is proportional to the intensity noise power that is contained within the detection spectral band. If this band is large enough to contain all the field-induced intensity noise spectrum, then the field-induced component of the current variance will be proportional to the output intensity variance Var[I']. The incoherent limit approximation provides a simple formula for this quantity.

To compute the output intensity variance, we use (9.49) and (10.64):

$$\mathrm{Var}[I'] = \int df\{K(f)N(f) + [L(f) - K(f)]N^{(G)}(f)\} \tag{10.71}$$

Applying the decomposition theorem to the right-hand side, we get

$$\mathrm{Var}[I'] = \{K(f)\}_f \mathrm{Var}[I] + [\{L(f)\}_f - \{K(f)\}_f]\mathrm{Var}[I^{(G)}] \tag{10.72}$$

where Var[I] and Var[$I^{(G)}$] are the intensity variances of the source and its Gaussian equivalent, respectively. Again, it is more convenient to deal with the *relative* rather than with the absolute intensity variance. Dividing (10.72) by $\langle I' \rangle^2$ and denoting the relative variance of the output field intensity by *RIV'*, we obtain

$$RIV' = \frac{\{K(f)\}_f}{\{T(\nu)\}_\nu^2} RIV + \frac{\{L(f)\}_f - \{K(f)\}_f}{\{T(\nu)\}_\nu^2} RIV^{(G)} \tag{10.73}$$

where *RIV* is the relative variance of the source intensity

$$RIV = \frac{\text{Var[I]}}{\langle I \rangle^2}$$

and $RIV^{(G)}$ is the relative variance of the equivalent Gaussian source intensity

$$RIV^{(G)} = \frac{\text{Var}[I^{(G)}]}{\langle I^{(G)} \rangle^2} \quad \text{(by definition } \langle I \rangle = \langle I^{(G)} \rangle)$$

By (9.51), the relative intensity variance of the Gaussian equivalent of a polarized source is 1.

Although we cannot offer a simple expression for $\{K(f)\}_f$, there exists a simple relation between $\{L(f)\}_f$ and the average network power transmission $\{T(\nu)\}_\nu$. From (10.63) we get

$$\{L(f)\}_f = \{|J(\nu)|^2 |J(\nu + f)|^2\}_{\nu,f}$$

Since only the second factor depends on f, we may write

$$\{L(f)\}_f = \{|J(\nu)|^2 \{|J(\nu + f)|^2\}_f\}_\nu$$

The integral of a periodic function over an interval with the length of the period does not depend on the location of that interval on the integration variable axis. Therefore,

$$\{|J(\nu + f)|^2\}_f = \{|J(\nu)|^2\}_\nu = \{T(\nu)\}_\nu \tag{10.74}$$

This leads us to

$$\{L(f)\}_f = \{T(\nu)\}_\nu^2 \tag{10.75}$$

We introduce now the *network variance coefficient NVC*:

$$NVC = \frac{\{K(f)\}_f}{\{T(\nu)\}_\nu^2} \tag{10.76}$$

The value of *NVC* lies always between 0 and 1. The output relative intensity variance can be now put in the form

$$RIV' = NVC \cdot RIV + (1 - NVC) \cdot RIV^{(G)} \tag{10.77}$$

Thus, in the incoherent limit, the output relative intensity variance is a simple linear function of the source and its equivalent Gaussian relative intensity variances. For a random-phase source, $RIV = 0$, and

$$RIV' = (1 - NVC) \cdot RIV^{(G)} \tag{10.78}$$

which is the smallest possible value that can be obtained, since, obviously, for any source, *RIV* is nonnegative. On the other hand, if the source is Gaussian, then $RIV = RIV^{(G)}$ and, by (10.77),

$$RIV' = RIV^{(G)} \tag{10.79}$$

This result implies that in the incoherent limit, a polarization-degenerate network does not affect the relative intensity variance of a Gaussian source. Finally, for a polarized source we have

$$RIV' = NVC \cdot RIV + 1 - NVC \tag{10.80}$$

since for a polarized source $RIV^{(G)} = 1$.

10.6 SUMMARY

The evaluation of the output intensity noise power spectrum is, in general, a difficult problem. This power spectrum depends on the fourth-order correlation function of the source, which is unavailable in most cases. In addition, even if this correlation function is known, the computation of the output intensity noise power spectrum involves the evaluation of a two-dimensional integral. Fortunately, in many cases of interest it is possible to use various approximations, which lead to significant simplifications.

One type of simplification may be classified as algebraic. These simplifications are invoked if the source is separable or the network is polarization degenerate. In particular, if both conditions occur, we may treat each polarization mode separately, thus replacing the original (matrix) problem with two independent scalar problems.

The general expressions are significantly simplified if Gaussian statistics are assumed for the source. This type of simplification may be regarded as being of statistical origin. However, its practical significance is quite limited, since Gaussian statistics is inadequate for the description of laser diodes, which are the most common sources for optical networks.

An important case which lends itself to a simple analytic treatment is the case of the random-phase source coupled to a dispersive waveguide. This case is of particular importance in long optical communication links equipped with nonregenerative amplifiers, where significant dispersion may be accumulated. Our treatment yields a simple analytical expression for the output intensity noise power spectrum, which qualitatively compares well with the experiment.

A very powerful and general simplification is introduced by the incoherent limit approximation. It is valid for any source-network combination satisfying the incoherent limit condition. In this approximation, the output intensity noise power spectrum is expressed in terms of the source intensity noise power spectrum and certain network characteristic functions. All these quantities are usually available, at least in an approximate form. In addition, the network and the source-dependent terms are factorized, which means that once the network-dependent terms are evaluated, the problem is essentially solved for any source. In the incoherent limit we have also derived simple expressions for the output intensity variance.

References

[1] Weissman, Y., "The Theory of Source-Induced Noise in Optical Networks," *J. Opt. Soc. Am.*, Vol. 7, No. 9, September 1990, pp. 1791-1800.

[2] Yamamoto, S., N. Edagawa, H. Taga, Y. Yoshida, and H. Wakabayashi, "Analysis of Laser Phase Noise to Intensity Noise Conversion by Chromatic Dispersion in Intensity Modulation and Direct Detection Optical-Fiber Transmission," *IEEE J. Lightwave Technol.*, Vol. 8, No. 11, November 1990, pp. 1716-1722.

Chapter 11
Analytic Methods for the Incoherent Limit

In the previous chapter we introduced the network characteristic functions for incoherent networks. In this chapter we develop some general methods for the computation of these functions and illustrate them with three examples.

The computation of the network characteristic functions involves the evaluation of an integral of a periodic function over the period interval. Since the function is regular and the integration interval is finite, practically any integration routine can handle the numerical aspect of this task. Indeed, in many cases a straightforward numerical integration may be the best way to proceed. However, in using the numerical approach, it is difficult to see the overall behavior as a function of certain network parameters that might be of interest. In addition, this approach may be too time consuming, especially if the characteristic functions are to be integrated again (for instance, in order to evaluate the effectiveness of a bandpass electronic filter). Whenever these shortcomings turn out to be crucial, it will be constructive to apply the analytical tools presented here.

The periodic functions that we deal with depend on the frequency ν only through the exponential factor $e^{2\pi i\nu T}$, where T is the common delay time of the network. In this case it is natural to resort to the complex residue calculus for the computation the frequency integrals. The integral over the frequency period interval Δf is converted to a contour integral over the unit circle in the complex plane, and the contour integral is expressed as the sum over all residues of the integrand that correspond to its poles inside this circle. Quite often these residues can be computed analytically, resulting in analytic expressions for the integrals. It is true, however, that as the number of the poles increases, these expressions become increasingly complicated.

For simplicity, throughout this chapter we will consider only characteristic functions, and not characteristic matrices. Therefore, the results presented here apply to polarization-degenerate networks only. The extension to the case of a polarized source or a polarizing network is trivial: one simply has to consider the corresponding matrix elements of the characteristic matrices instead of the characteristic functions.

The extension to the general case involves some algebra, but otherwise is completely straightforward.

11.1 APPLICATION OF THE RESIDUE CALCULUS FOR THE CALCULATION OF THE AVERAGING INTEGRAL

Let us consider a function $F(\nu)$ that depends on ν only through the exponential $e^{2\pi i\nu T}$. Such a function is obviously periodic in ν, with a period of $\Delta f = 1/T$. Our task is to compute the integral

$$I = \{F(\nu)\}_\nu = \frac{1}{\Delta f} \int_{[\Delta f]} d\nu F(\nu) \tag{11.1}$$

We now derive from $F(\nu)$ a new function $f(z)$ by replacing each exponential factor $e^{2\pi i\nu T}$ with a complex variable z:

$$f(z) \equiv F(\nu), \quad z = e^{2\pi i\nu T} \tag{11.2}$$

In our case it will be possible to regard $f(z)$ as an analytic function in the complex plane, excluding a finite number of points which are its poles. Using the complex variable z, the integral (11.1) is converted to a contour integral over the unit circle $|z| = 1$:

$$I = \frac{1}{2\pi i} \oint \frac{dz}{z} f(z) \tag{11.3}$$

since

$$d\nu = \frac{\Delta f}{2\pi i} \frac{dz}{z}$$

Suppose that $f(z)$ has M poles $z_1, \ldots, z_M$ inside the unit circle that are different from 0. According to the Cauchy's residue theorem [1], we have

$$I = \mathrm{Res}[\frac{f(z)}{z}, 0] + \sum_k \mathrm{Res}[\frac{f(z)}{z}, z_k] \tag{11.4}$$

where we denoted the residue of a function $q(z)$ at a pole z_k with $\mathrm{Res}[q(z), z_k]$. The general formula for the computation of residues is [1]

$$\mathrm{Res}[f(z),z_k] = \frac{1}{(n-1)!} \lim_{z \to z_k} \frac{d^{n-1}}{dz^{n-1}} [(z - z_k)^n f(z)] \tag{11.5}$$

where n is the order of the pole. For a simple pole, $n = 1$, and the general formula (11.5) reduces to

$$\mathrm{Res}[f(z),z_k] = \lim_{z \to z_k} [(z - z_k) f(z)] \tag{11.6}$$

In particular, it follows that if $P(z)$ has a simple pole at $z = z_0$, and $Q(z)$ is regular at this point, then

$$\mathrm{Res}[P(z)Q(z),z_0] = Q(z_0)\mathrm{Res}[P(z),z_0] \tag{11.7}$$

Therefore, if all poles of $f(z)$ inside the unit circle are simple and different from 0, then

$$I = f(0) + \sum_k \frac{\mathrm{Res}[f(z),z_k]}{z_k} \tag{11.8}$$

If $z = 0$ is a simple pole of $f(z)$, then it is a second-order pole of $\frac{f(z)}{z}$. Using (11.5), we find for this case that

$$\mathrm{Res}[\frac{f(z)}{z},0] = \lim_{z \to 0} \frac{d}{dz}[z f(z)] \tag{11.9}$$

Consequently, in the case that $f(z)$ has a simple pole at $z = 0$, (11.4) has to be replaced by

$$I = \lim_{z \to 0} \frac{d}{dz}[z f(z)] + \sum_k \frac{\mathrm{Res}[f(z),z_k]}{z_k}$$

11.2 COMPUTATION OF THE NETWORK CHARACTERISTIC FUNCTIONS

11.2.1 The *K* Characteristic Function

The K characteristic function (Eq. (10.62)) is given by

$$K(f) = |I(f)|^2 \tag{11.10}$$

where

$$I(f) = \frac{1}{\Delta f} \int_{[\Delta f]} d\nu J^*(\nu + f) J(\nu) \tag{11.11}$$

and $J(\nu)$ is the Jones (transfer) function of the network. As we deal here with the incoherent limit, it is assumed that $J(\nu)$ depends on ν only through the factor $e^{2\pi i\nu T}$. It is therefore possible to apply the method of the complex contour integration described above to evaluate this integral.

We start by introducing a function $j(z)$ derived from $J(\nu)$ by the replacement of $e^{2\pi i\nu T}$ with z, as we have done in (11.2) above, and regard it as a complex analytic function in the complex plane. This operation is straightforward. However, we also have to replace $J^*(\nu)$ with an appropriate analytic function. For this purpose we might have used $[j(z)]^*$. This choice, though perfectly valid, would have led us to a dead end, since $[j(z)]^*$ contains z^*, and the conventional complex contour integration methods cannot be applied to integrands that depend on both z and z^*. To overcome this difficulty, we note that in $J^*(\nu)$ each $e^{2\pi i\nu T}$ factor is replaced by its complex conjugate, i.e., $e^{-2\pi i\nu T}$, which is also $1/z$. Thus, in converting (11.11) to a contour integral, we make the substitutions

$$J(\nu) \rightarrow j(z) \tag{11.12}$$

and

$$J^*(\nu) \rightarrow j^*\left(\frac{1}{z}\right) \tag{11.13}$$

In (11.13) we have introduced a new notation, $j^*(z)$, which may be defined as

$$j^*(z) \equiv [j(z^*)]^* \tag{11.14}$$

To illustrate the new notation, let us consider the simple case of $j(z) = A + Bz$:

$$[j(z)]^* = A^* + B^* z^*$$

However,

$$j^*(z) = A^* + B^*z$$

Thus, in computing $j^*(z)$, we exclude the argument z from the complex conjugation. Obviously, the distinction between $[F(x)]^*$ and $F^*(x)$ is unnecessary for complex functions of *real* arguments like $J(\nu)$.

To find the analytic function corresponding to $J^*(\nu + f)$ we note that

$$e^{2\pi i(\nu+f)T} = e^{2\pi i f T}z$$

so that

$$J^*(\nu + f) = j^*(\frac{1}{e^{2\pi i f T}z}) \tag{11.15}$$

To simplify the notation, we introduce the angle φ:

$$\varphi = 2\pi f T \tag{11.16}$$

We can now cast (11.11) in the form

$$I(\varphi) = \frac{1}{2\pi i}\oint \frac{dz}{z} j^*(\frac{1}{e^{i\varphi}z}) j(z) \tag{11.17}$$

Let us assume that $j(z)$ has M finite poles $z_1, ..., z_M$. We claim that in a passive system, all poles of $j(z)$ lie outside the unit circle; i.e., $|z_k| > 1$ for all k. This claim is justified by the following physical argument. The Jones function $J(\nu)$ relates the input and the output Jones vectors $\boldsymbol{A}(\nu)$ and $\boldsymbol{A}'(\nu)$ (Eq. (2.15)):

$$\boldsymbol{A}'(\nu) = J(\nu)\boldsymbol{A}(\nu) = j[z(\nu)]\boldsymbol{A}(\nu)$$

In a passive system, the output power cannot exceed the input power; i.e.,

$$|\boldsymbol{A}'(\nu)|^2 \le |\boldsymbol{A}(\nu)|^2 \tag{11.18}$$

The factors $e^{2\pi i \nu T}$ in $J(\nu)$ represent the transmission of the waveguides. We assumed that the waveguides are ideally lossless; in real waveguides the transmission is $e^{2\pi i \nu T - \alpha}$, where α is a positive real constant representing the loss. Thus, the value of z corresponding to a real waveguide lies inside the unit circle. Suppose now that z_0 is a pole of $j(z)$ internal to the unit circle. By "turning on" a waveguide loss $\alpha = -\ln|z_0|$, we obtain a system with an infinite value of $J(\nu)$ at a frequency ν_0 determined by $\arg(z_0) = 2\pi\nu_0 T$. At this frequency, (11.18) cannot be satisfied. Since (11.18) must be satisfied in a passive system for all frequencies and regardless of waveguide losses, we conclude that in such a system $j(z)$ cannot have a pole internal to the unit circle. Throughout this chapter, we will limit ourselves to passive systems.

Since all the poles of $j(z)$ are located outside the unit circle, all poles of the integrand in (11.17) that are internal to the unit circle come from $j^*(\frac{1}{e^{i\varphi}z})$ and $1/z$. There are, altogether, $M + 1$ such poles:

$$u_k = \frac{1}{e^{i\varphi}z_k^*}, \quad k = 1, ..., M$$

and

$$u_0 = 0$$

We denote the residues of $j(z)$ with

$$R_k = \mathrm{Res}[j(z), z_k]$$

The expansion of $j(z)$ in Laurant series [1] around z_k is

$$j(z) = \frac{R_k}{z - z_k} + j_k(z) \tag{11.19}$$

where $j_k(z)$ contains the rest of the expansion. From this expansion we obtain

$$j^*(\frac{1}{e^{i\varphi}z}) = -\frac{z}{z_k^*}\frac{R_k^*}{z - \frac{1}{e^{i\varphi}z_k^*}} + j_k^*(\frac{1}{e^{i\varphi}z})$$

so that

$$\mathrm{Res}\left[j^*(\frac{1}{e^{i\varphi}z}), u_k\right] = -\frac{R_k^*}{e^{i\varphi}z_k^{*2}}, \quad k = 1, ..., M. \tag{11.20}$$

Using now (11.7), we get

$$\mathrm{Res}\left[j^*(\frac{1}{e^{i\varphi}z})j(z)\frac{1}{z}, u_k\right] = -\frac{R_k^*}{z_k^*}j(\frac{1}{e^{i\varphi}z_k^*}), \quad k = 1, ..., M. \tag{11.21}$$

Using Cauchy's theorem, we can represent $I(f)$ as follows:

$$I(f) = R_0 - \sum_{k=1}^{M} \frac{R_k^*}{z_k^*} j\left(\frac{1}{e^{i\varphi} z_k^*}\right) \tag{11.22}$$

where

$$R_0 = \mathrm{Res}\left[j^*\left(\frac{1}{e^{i\varphi} z}\right) j(z) \frac{1}{z}\ , 0 \right] \tag{11.23}$$

Finally, it is worthwhile to note that

$$\{T(\nu)\}_\nu = I(0) = R_0 - \sum_{k=1}^{M} \frac{R_k^*}{z_k^*} j\left(\frac{1}{z_k^*}\right) \tag{11.24}$$

This is a simple formula for the computation of the average power transmission of the network.

11.2.2 The L Characteristic Function

The computation of the L characteristic function is more complex than the computation of the K characteristic function presented above, but it rests on the same general ideas. Writing explicitly the averaging integral in the definition of $L(f)$, we get

$$L(f) = \frac{1}{\Delta f} \int_{[\Delta f]} d\nu |J(\nu + f) J(\nu)|^2 \tag{11.25}$$

In terms of $j(z)$ and φ, this equation reads

$$L(\varphi) = \frac{1}{2\pi i} \oint \frac{dz}{z} j(e^{i\varphi} z) j^*\left(\frac{1}{e^{i\varphi} z}\right) j(z) j^*\left(\frac{1}{z}\right) \tag{11.26}$$

The integrand in this integral has $2M + 1$ poles that are internal to the unit circle: $u_k = 1/(e^{i\varphi} z_k^*)$, $v_k = 1/z_k^*$, and 0. The residue of $j^*(\frac{1}{z})$ at v_k is computed in a manner similar to the calculation preceding (11.20). From (11.19) we get

$$j^*\left(\frac{1}{z}\right) = -\frac{z}{z_k^*} \frac{R_k^*}{z - \frac{1}{z_k^*}} + j_k^*\left(\frac{1}{z}\right)$$

so that

$$\mathrm{Res}\left[j^*\left(\frac{1}{z}\right), v_k \right] = -\frac{R_k}{z_k^{*2}} \tag{11.27}$$

Finally, using (11.27), (11.20), and (11.7), we obtain

$$\mathrm{Res}\left[j(e^{i\varphi}z)j^*(\frac{1}{e^{i\varphi}z})j(z)j^*(\frac{1}{z})\frac{1}{z}, u_k\right] = -\frac{R_k^*}{z_k^*}j(\frac{1}{z_k^*})j(\frac{1}{e^{i\varphi}z_k^*})j^*(e^{i\varphi}z_k^*) \quad (11.28)$$

and

$$\mathrm{Res}\left[j(e^{i\varphi}z)j^*(\frac{1}{e^{i\varphi}z})j(z)j^*(\frac{1}{z})\frac{1}{z}, v_k\right] = -\frac{R_k^*}{z_k^*}j(\frac{1}{z_k^*})j(\frac{e^{i\varphi}}{z_k^*})j^*(e^{-i\varphi}z_k^*) \quad (11.29)$$

Using Cauchy's residue theorem, we obtain the following expression for $L(\varphi)$:

$$L(\varphi) = P_0 - \sum_k \frac{R_k^*}{z_k^*}j(\frac{1}{z_k^*})[\alpha_k(\varphi) + \alpha_k(-\varphi)] \quad (11.30)$$

where

$$\alpha_k(\varphi) = j(\frac{e^{i\varphi}}{z_k^*})j^*(e^{-i\varphi}z_k^*) \quad (11.31)$$

and

$$P_0 = \mathrm{Res}\left[j(e^{i\varphi}z)j^*(\frac{1}{e^{i\varphi}z})j(z)j^*(\frac{1}{z})\frac{1}{z}, 0\right] \quad (11.32)$$

In deriving (11.30), we have overlooked a small complication. For $\varphi = 0$, the poles v_k and u_k coincide and become of second order. Indeed, for $\varphi = 0$ the right-hand side of (11.31) becomes singular. For physical reasons, $L(\varphi)$ itself does not become singular, and, therefore, taking the limit $\varphi \to 0$ in (11.30) should yield a finite result. A similar problem may occur if there are two poles, say z_k and z_j, such that $|z_k| = |z_j|$. In this case, there is a certain angle φ_0 such that v_k coincides with u_k, creating together a second-order pole. Again, the correct value of $L(\varphi)$ will be obtained if in (11.30) the limit $\varphi \to \varphi_0$ is taken.

11.3 THE GUIDED-WAVE FABRY-PEROT INTERFEROMETER (EXAMPLE)

The incoherent limit of the guided-wave Fabry-Perot interferometer has been thoroughly investigated, for both random-phase [2-4] and Gaussian sources [5,6]. Below, we use the frequency-domain representation and the contour-integration method to derive the noise factors for this network. The present treatment differs from the traditional time-domain approach used in [2-5]. Needless to say, both approaches are equally valid and yield identical results.

11.3.1 The *K* Characteristic Function

The Jones function $J(\nu)$ of the guided-wave Fabry-Perot interferometer was derived in (5.5):

$$J(\nu) = \frac{e^{2\pi i\nu\tau}t^2}{1 - e^{4\pi i\nu\tau}r^2}$$

where the parameters r and t denote the (field) reflectance and transmittance of the two mirrors, respectively. As we have shown in the previous chapter, the exponential factor in the nominator can be disregarded, and defining $z = e^{4\pi i\nu\tau}$ we get the following $j(z)$ function for the Fabry-Perot interferometer:

$$j(z) = \frac{t^2}{1 - r^2 z}$$

This function contains only one pole: $z_1 = 1/r^2$. For a passive mirror, $|r^2| < 1$, so that z_1 lies outside the unit circle. A straightforward calculation yields

$$R_1 = -\frac{t^2}{r^2} \tag{11.33}$$

and

$$j\left(\frac{1}{e^{i\varphi}z_1{}^*}\right) = \frac{t^2}{1 - |r|^4 e^{-i\varphi}} \tag{11.34}$$

It is easy to verify that, in this case,

$$\operatorname{Res}\left[j^*\left(\frac{1}{e^{i\varphi}z}\right)j(z)\frac{1}{z}\ ,0\right] = 0$$

so that $R_0 = 0$. Using (11.22) we get

$$I(\varphi) = \frac{|t|^4}{1 - |r|^4 e^{-i\varphi}} \tag{11.35}$$

and finally

$$K(\varphi) = |I(\varphi)|^2 = \frac{|t|^8}{1 - 2|r|^4\cos\varphi + |r|^8} \tag{11.36}$$

From (11.35) we get the average power transmission of the Fabry-Perot interferometer:

$$\{T(\nu)\}_\nu = I(0) = \frac{|t|^4}{1 - |r|^4} \tag{11.37}$$

11.3.2 The L Characteristic Function

We turn now to the computation of the L characteristic function. Again, it is easy to verify that, in the present case,

$$\mathrm{Res}\left[j(e^{i\varphi}z)j^*(\frac{1}{e^{i\varphi}z})j(z)j^*(\frac{1}{z})\frac{1}{z}\ ,0\right] = 0$$

i.e., $P_0 = 0$. From (11.33) and (11.34) we get

$$\frac{R_k^*}{z_k^*}j(\frac{1}{z_k^*}) = \frac{|t|^4}{1 - |r|^4} \tag{11.38}$$

Since, in the Fabry-Perot case, $j(z)$ possesses only one pole, all that is left is to compute the angle function $\alpha_1(\varphi)$. Using (11.31) we find

$$\alpha_1(\varphi) = \frac{1}{(1 - |r|^4 e^{i\varphi})(1 - e^{-i\varphi})} \tag{11.39}$$

Using (11.30) we obtain the final expression for $L(\varphi)$:

$$L(\varphi) = \frac{1 + |r|^4}{1 - |r|^4}\frac{|t|^8}{1 + |r|^8 - 2|r|^4\cos\varphi} = \frac{1 + |r|^4}{1 - |r|^4}K(\varphi) \tag{11.40}$$

11.3.3 The Fabry-Perot Noise Factors

The Fabry-Perot interferometer noise factors $NNF^{(G)}(\varphi)$ and $NNF^{(rp)}(\varphi)$ are derived from (10.66) and (10.67). Using the corresponding expressions for the characteristic functions $K(\varphi)$ and $L(\varphi)$ derived above, by a simple substitution we obtain

$$NNF^{(G)}(\varphi) = \frac{1 - |r|^8}{1 + |r|^8 - 2|r|^4\cos\varphi} \tag{11.41}$$

and

$$NNF^{(rp)}(\varphi) = \frac{2|r|^4(1 - |r|^4)}{1 + |r|^8 - 2|r|^4\cos\varphi} \tag{11.42}$$

In the Fabry-Perot case, there is a simple proportionality relation between the two noise factors:

$$NNF^{(rp)}(\varphi) = \frac{2|r|^4}{1 + |r|^4} NNF^{(G)}(\varphi)$$

It is interesting to note that both noise factors do not depend on t. Since $|r| < 1$, $NNF^{(rp)}(\varphi)$ is smaller than $NNF^{(G)}(\varphi)$. This conforms to our earlier conclusion that for a source with a given lineshape, the random-phase statistics yields the smallest output intensity noise. We have plotted $NNF^{(rp)}(\varphi)$ as a function of φ and for several values of $|r|^2$ in Figure 11.1. The analytical expressions for $NNF^{(rp)}$ and $NNF^{(G)}$ have been derived already using a time-domain approach [4,5].

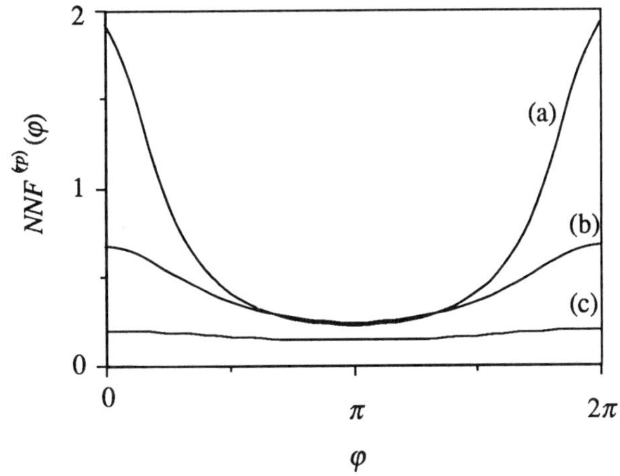

Figure 11.1 The network noise factor $NNF^{(rp)}(\varphi)$ for the Fabry-Perot system: (a) $|r|^2 = 0.7$; (b) $|r|^2 = 0.5$; (c) $|r|^2 = 0.3$.

Figure 11.2 illustrates how the output relative intensity noise may look in practice. We have assumed that the relative intensity noise power spectrum $RIN^{(G)}(f)$ of the equivalent Gaussian source is in the form (Eq. (10.36)):

$$RIN^{(G)}(f) = \frac{2\gamma}{\gamma^2 + (2\pi f)^2}$$

By (10.69), the output relative intensity noise for a random-phase source will be

$$RIN'(f) = \frac{2\gamma}{\gamma^2 + (2\pi f)^2} NNF^{(rp)}(f) \tag{11.43}$$

In Figure 11.2 we plotted $\gamma RIN'(f)$ as a function of $f/\Delta f$ for the case

$$\frac{\Delta f}{\gamma} = 0.1 \tag{11.44}$$

which is compatible with the incoherent limit requirement.

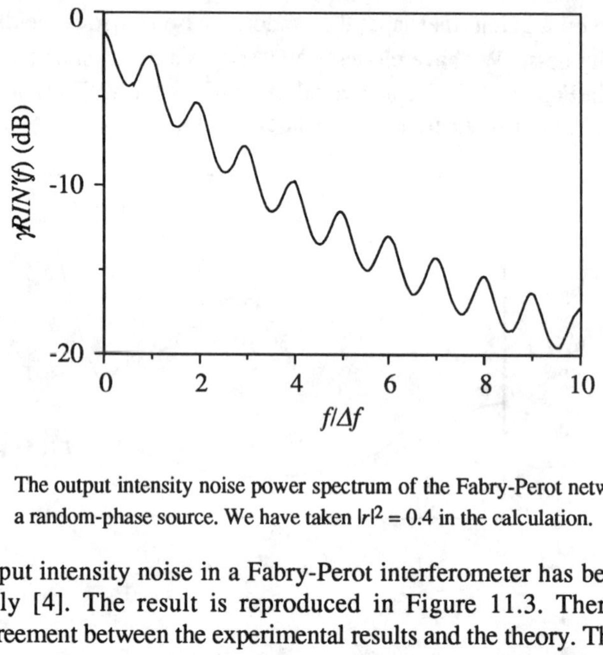

Figure 11.2 The output intensity noise power spectrum of the Fabry-Perot network coupled to a random-phase source. We have taken $|r|^2 = 0.4$ in the calculation.

The output intensity noise in a Fabry-Perot interferometer has been measured experimentally [4]. The result is reproduced in Figure 11.3. There is a good qualitative agreement between the experimental results and the theory. The maxima of the experimental noise spectrum appear at integer values of $\Delta f/f$, and the peak-to-peak modulation of the noise spectrum curve is approximately 2 dB, as in Figure 11.2.

11.3.4 The Variance Coefficient for the Fabry-Perot Interferometer

To derive the network variance coefficient, we need the frequency average of the K characteristic function (10.76). Naturally, we will use the complex contour integration method to compute this average. We write $K(\varphi)$ as follows:

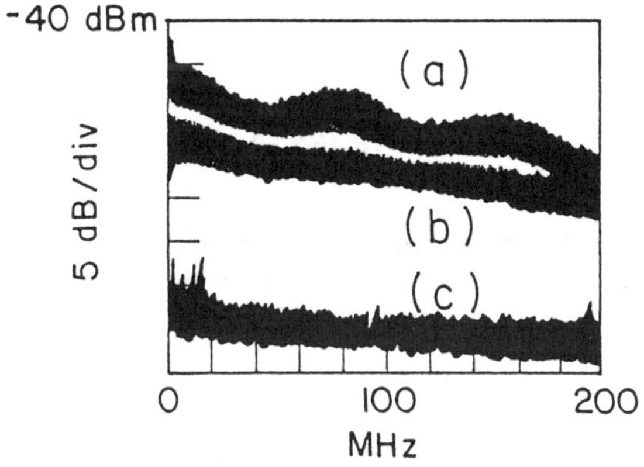

Figure 11.3 The output intensity noise spectrum of a Fabry-Perot interferometer: (a) tuned cavity; (b) untuned cavity; (c) laser off. The source was a multimode laser diode, and the mirror's intensity reflectance $|r|^2$ had a value of 0.38. The cavity length was 375 cm [4] (© JOSA 1987).

$$K(\varphi) = \frac{|t|^8}{1 - |r|^4(e^{i\varphi} + e^{-i\varphi}) + |r|^8}$$

By introducing $z = e^{i\varphi}$ we derive a complex function $k(z)$:

$$k(z) = \frac{|t|^8}{1 - |r|^4(z + \frac{1}{z}) + |r|^8}$$

This function allows us to express the frequency average as a complex contour integral over the unity circle:

$$\{K(f)\}_f = \frac{1}{2\pi i} \oint \frac{dz}{z}\, k(z)$$

The integrand has only one pole internal to the unity circle at the point $|r|^4$. Elementary application of the residue calculus yields

$$\{K(f)\}_f = \frac{|t|^8}{1 - |r|^8}$$

Using the value of $\{T(\nu)\}_\nu$ calculated above (Eq. (11.37)), we get

$$NVC = \frac{1 - |r|^4}{1 + |r|^4}$$

11.4 THE RECIRCULATING LOOP (EXAMPLE)

The incoherent limit of the recirculating loop has also been studied extensively [3,4,6]. As a second example of the contour integration method, we present below a treatment of the simple case of the lossless recirculating loop.

11.4.1 The Characteristic Functions

The Jones function of the recirculating loop (5.9) can be cast in the form

$$J(\nu) = \frac{e^{-i\phi}\cos\theta - e^{2\pi i\nu\tau}}{1 - e^{i\phi + 2\pi i\nu\tau}\cos\theta}$$

The corresponding j function is therefore

$$j(z) = \frac{e^{-i\phi}\cos\theta - z}{1 - e^{i\phi}z\cos\theta} \tag{11.45}$$

This function has only one pole z_1. The values of z_1 and the corresponding residue are

$$z_1 = \frac{1}{e^{i\phi}\cos\theta}$$

$$R_1 = e^{-i\phi}\,\frac{\sin^2\theta}{\cos\theta}$$

It is easy to verify that, in this case,

$$\mathrm{Res}\left[j^*\left(\frac{1}{e^{i\varphi}z}\right)j(z)\frac{1}{z}\ ,0\right] = 1 \tag{11.46}$$

To compute $I(\varphi)$, we need to evaluate $j\left(\frac{1}{e^{i\varphi}z_1{}^*}\right)$. From (11.45) we obtain

$$j\left(\frac{1}{e^{i\varphi}z_1{}^*}\right) = \frac{e^{-i\phi}\cos\theta(1 - e^{-i\varphi})}{1 - e^{-i\varphi}\cos^2\theta} \tag{11.47}$$

We are ready now to evaluate $I(\varphi)$. Using (11.22) we get

$$I(\varphi) = \frac{\mu + (1 - 2\mu)e^{-i\varphi}}{1 - \mu e^{-i\varphi}} \tag{11.48}$$

where we have introduced

$$\mu = \cos^2\theta$$

The calculation of $K(\varphi)$ is now straightforward:

$$K(\varphi) = |I(\varphi)|^2 = \frac{\mu^2 + 2\mu(1 - 2\mu)\cos\varphi + (1 - 2\mu)^2}{1 - 2\mu\cos\varphi + \mu^2} \tag{11.49}$$

From (11.24) it follows that, in the recirculating loop case,

$$\{T(\nu)\}_\nu = 1 \tag{11.50}$$

In fact, in the recirculating loop case, $T(\nu) = 1$ independently of ν, as we noted already in the discussion following (5.9). The reason for this is, of course, our neglect of the losses in the coupler and the waveguide, resulting in an ideally lossless system.

We turn now to the computation of $L(\varphi)$. From (11.40) it follows that

$$j\left(\frac{1}{z_1{}^*}\right) = 0$$

Consequently,

$$L(\varphi) = 1$$

11.4.2 The Recirculating Loop Noise Factors

The noise factors for the recirculating loop network are given by

$$NNF^{(G)} = 1 \tag{11.51}$$

and

$$NNF^{(rp)}(\varphi) = 1 - \frac{\mu^2 + 2\mu(1 - 2\mu)\cos\varphi + (1 - 2\mu)^2}{1 - 2\mu\cos\varphi + \mu^2} \tag{11.52}$$

Referring to Figure 5.4(a), the parameter μ gives the fraction of power that is transmitted directly from the source to the detector. In the limit $\mu \to 1$, the loop becomes actually decoupled; in the limit $\mu \to 0$, all the power is cross-coupled and reaches the detector after a single pass through the loop. In both of these cases the source statistics remains unaffected, and correspondingly $NNF^{(rp)}(\varphi) \to 1$. Another peculiarity of the recirculating loop system is the fact that the intensity noise of a Gaussian source remains unaffected by it. This can be explained as follows. Since the power transmission of the ideal recirculating loop is unity, the output power spectrum is identical to the input power spectrum. We have seen, however, that the intensity noise of a separable Gaussian field is completely determined by its (polarization mode) lineshapes. Therefore, if the lineshapes are equal, so are the corresponding intensity noise power spectra (a linear network does not impair the Gaussian statistics, since a linear combination of Gaussian random variables is also a Gaussian random variable). It follows, therefore, that in real systems, where losses are inevitable, deviations of $L(\varphi)$ from unity exist. These deviations, however, are rather weak if the losses are small. In Figure 11.4 we show several plots of $NNF^{(rp)}(\varphi)$ for three representative values of μ. The analytic expression for $NNF^{(rp)}(\varphi)$ in the recirculating loop case has also been derived using the time-domain approach [4,7].

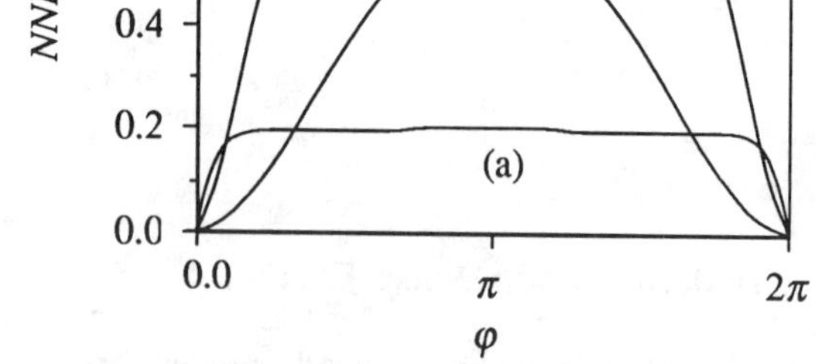

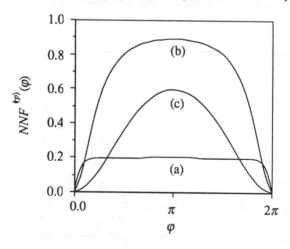

Figure 11.4 The recirculating loop noise factor $NNF^{(rp)}(\varphi)$: (a) $\mu = 0.1$; (b) $\mu = 0.5$; (c) $\mu = 0.9$.

To illustrate how the output intensity noise may look in practice, we have plotted in Figure 11.5 the output relative intensity noise power spectrum for $\mu = 0.5$, using (11.43) and (11.44). Since $NNF^{(rp)}$ vanishes for integer values of $f/\Delta f$, the logarithm of $RIN'(f)$ exhibits singularities at these points. To avoid these singularities, we have superimposed $RIN'(f)$ in Figure 11.5 on a flat background noise. This operation can also be motivated physically, since at frequencies in which the field-induced noise power vanishes, the system noise will be determined by a different noise mechanism, like the shot noise.

In comparing Figures 11.2 and 11.5, we note a remarkable difference. In the Fabry-Perot case, the output relative intensity noise exhibits maxima for integer values of $f/\Delta f$. In the recirculating loop case, the behavior of the output relative intensity noise is reversed: for integer values of $f/\Delta f$ deep minima are obtained. Thus for different network topologies, qualitatively different output intensity noise power spectra are generated.

The output intensity noise power spectrum of the recirculating loop network has been measured experimentally [7]. A typical experimental result is shown in Figure 11.6. The loop length in the experiment was 27 cm, corresponding to $\Delta f = 740$ MHz.

11.4.3 The Recirculating Loop Variance Coefficient

To compute the variance coefficient for the recirculating loop network, we proceed in exactly the same manner as we did in subsection (11.3.4) above. We write the K characteristic function (11.49) as follows:

$$K(\varphi) = \frac{\mu^2 + \mu(1 - 2\mu)(e^{i\varphi} + e^{-i\varphi}) + (1 - 2\mu)^2}{1 - \mu(e^{i\varphi} + e^{-i\varphi}) + \mu^2}$$

We now substitute z for $e^{i\varphi}$, and denote the resulting function with $k(z)$:

$$k(z) = \frac{\mu^2 + \mu(1 - 2\mu)(z + 1/z) + (1 - 2\mu)^2}{1 - \mu(z + 1/z) + \mu^2}$$

We now have

$$\{K(f)\}_f = \frac{1}{2\pi i}\oint \frac{dz}{z}\,\frac{\mu^2 + \mu(1 - 2\mu)(z + 1/z) + (1 - 2\mu)^2}{1 - \mu(z + 1/z) + \mu^2}$$

The integrand in this integral has only two poles internal to the unit circle: $z = 0$ and $z = \mu$. From elementary residue calculus, we obtain

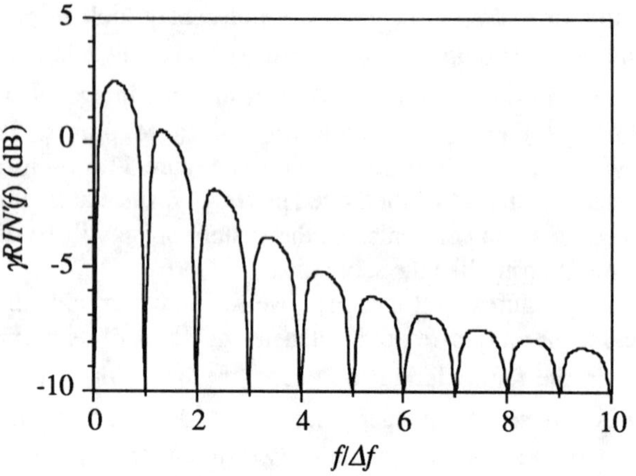

Figure 11.5 The relative output intensity noise of a recirculating loop with $\mu = 0.5$ coupled to a random-phase source. The noise spectrum is superimposed on a flat background.

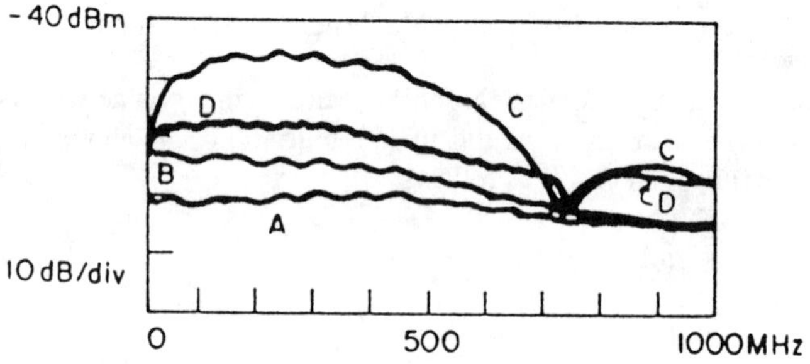

Figure 11.6 Experimental output intensity noise power spectrum of a recirculating loop: (a) laser off; (b) the source intensity noise; (d) and (c) the output intensity noise power spectrum with different values of μ [7] (© 1985 IEEE).

$$\mathrm{Res}\left[\frac{k(z)}{z}, \mu\right] = \frac{2(1-\mu)^2}{1+\mu}$$

and

$$\mathrm{Res}\left[\frac{k(z)}{z}, 0\right] = 2\mu - 1$$

Since for the recirculating loop $\{T(\nu)\}_\nu = 1$, we can write

$$NVC = \frac{4\mu^2 - 3\mu + 1}{1 + \mu} \tag{11.53}$$

We have plotted in Figure 11.7 the recirculating loop variance coefficient as a function of μ.

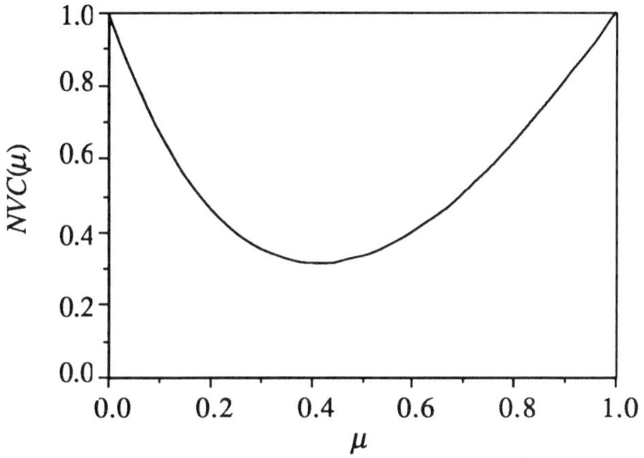

Figure 11.7 The variance factor for the recirculating loop network as a function of the directly transmitted power fraction μ.

11.5 THE GUIDED-WAVE MACH-ZEHNDER INTERFEROMETER (EXAMPLE)

11.5.1 The Characteristic Functions

The Jones function of the Mach-Zehnder interferometer is given by (Eq. (5.10))

$$J(\nu) = A\exp(2\pi i\nu\tau_1) + B\exp(2\pi i\nu\tau_2) \tag{11.54}$$

where A and B are complex constants. The Mach-Zehnder network differs from the two networks that were considered above in that it contains *two* delay times τ_1 and τ_2 instead of just one. Our general recipe calls for the determination of the common delay time T, such that $\tau_1 = n_1T$ and $\tau_2 = n_2T$. However, in the simple case of the

Mach-Zehnder network there is a shortcut. To avoid the necessity of introducing the common delay time, we cast the Jones function in the form

$$J(\nu) = \exp(2\pi i \nu \tau_2)(Ae^{2\pi i \nu \Delta\tau} + B) \tag{11.55}$$

where $\Delta\tau = \tau_1 - \tau_2$. As we explained in the previous chapter, this Jones function may be replaced by

$$J(\nu) = Ae^{2\pi i \nu \Delta\tau} + B \tag{11.56}$$

without affecting the relevant network characteristics. Therefore, for the present purposes, the Mach-Zehnder network may be regarded as a network with a single delay time $\Delta\tau$. Obviously, for a Jones function as simple as (11.56), it is possible to compute the characteristic functions directly, without invoking the contour integration method. However, in order to provide yet another example for this method, we will apply it to this case as well.

Proceeding as usual, we introduce $z = e^{2\pi i \nu \Delta\tau}$ and obtain

$$j(z) = Az + B \tag{11.57}$$

This function does not have any finite poles. Consequently, the pole sums in (11.23) and (11.30) vanish, and we are left in this case with

$$K(\varphi) = |R_0(\varphi)|^2 \tag{11.58}$$

and

$$L(\varphi) = P_0(\varphi) \tag{11.59}$$

Both $R_0(\varphi)$ and $P_0(\varphi)$ are computed using (11.9). Substitution of $Az + B$ for $j(z)$ in (11.23) and (11.32) yields

$$R_0(\varphi) = \lim_{z\to 0}\frac{d}{dz}[z(Az + B)(\frac{A^*}{e^{i\varphi}z} + B^*)] = |A|^2e^{-i\varphi} + |B|^2 \tag{11.60}$$

and

$$\begin{aligned} P_0(\varphi) &= \lim_{z\to 0}\frac{d}{dz}[z(Az + B)(\frac{A^*}{z} + B^*)(Ae^{i\varphi}z + B)(\frac{A^*}{e^{i\varphi}z} + B^*)] \\ &= (|A|^2 + |B|^2)^2 + 2|AB|^2\cos\varphi \end{aligned} \tag{11.61}$$

The K characteristic function is therefore given by

$$K(\varphi) = |R_0(\varphi)|^2 = |A|^4 + |B|^4 + 2|AB|^2\cos\varphi \tag{11.62}$$

and the L characteristic function is simply $P_0(\varphi)$. The average power transmission $\{T(\nu)\}_\nu$ is given by

$$\{T(\nu)\}_\nu = R_0(0) = |A|^2 + |B|^2 \tag{11.63}$$

11.5.2 The Mach-Zehnder Noise Factors

The noise factors $NNF^{(G)}(\varphi)$ and $NNF^{(rp)}(\varphi)$ for the Mach-Zehnder case are given by

$$NNF^{(G)}(\varphi) = 1 + \frac{2|AB|^2\cos\varphi}{(|A|^2 + |B|^2)^2} \tag{11.64}$$

and

$$NNF^{(rp)}(\varphi) = \frac{2|AB|^2}{(|A|^2 + |B|^2)^2} \tag{11.65}$$

This noise factor is constant. Therefore, the output intensity noise power spectrum of a Mach-Zehnder network coupled to a random-phase source is simply proportional to the intensity noise power spectrum of the equivalent Gaussian source.

Finally, let us compute the Mach-Zehnder network variance coefficient. From (11.62) it follows that

$$\{K(f)\}_f = |A|^4 + |B|^4$$

In view of (11.63), the variance coefficient in this case is given by

$$NVC = \frac{|A|^4 + |B|^4}{(|A|^2 + |B|^2)^2} \tag{11.66}$$

11.6 SUMMARY

In this chapter we have developed general methods for the analytic evaluation of the network characteristic functions $K(f)$ and $L(f)$. These methods are based on the fact that the integral of a periodic function over the period interval may be converted to an integral over the unit circle in the complex plane. This conversion is accomplished by an appropriate change of the integration variable. Such integrals may be evaluated by the Cauchy residue theorem. Using this theorem, we expressed the network characteristic functions in terms of the network Jones function poles and residues. These expressions simplify the analysis, since poles and residues normally are easier to evaluate than integrals.

The technique has been illustrated by treating three simple networks: the Fabry-Perot, the recirculating loop, and the Mach-Zehnder. All these networks have been treated previously, using the time-domain approach. While both the time-domain and the present approaches are mathematically equivalent, the present approach is advantageous in being able to treat multipole networks with relative ease.

The qualitative features of the output intensity noise power spectrum depend on both the source statistics and the network topology. We have seen, for instance, that, in the case of the random-phase source coupled to a Fabry-Perot interferometer, integer values of $f/\Delta f$ correspond to maxima in the spectrum. On the other hand, if a random-phase source is coupled to a recirculating loop, the same points become deep minima. Both these predictions have been confirmed experimentally. In the case of the Gaussian source coupled to a recirculating loop and the case of the random-phase source coupled to a Mach-Zehnder interferometer, the output power spectrum is flat, in the sense that it does not contain any spectral features on the scale of Δf.

References

[1] Hille, E., *Analytic Function Theory*, Vol. 1, Second Edition, New York: Chelsea Publishing Company, 1973.

[2] Armstrong, J. A., "Theory of Interferometric Analysis of Laser Phase Noise," *J. Opt. Soc. Am.*, Vol. 56, No. 8, August 1966, pp. 1024-1031.

[3] Moslehi, B., "Analysis of Optical Phase noise in Fiber-Optic Systems Employing a Laser Source with Arbitrary Coherence Time," *IEEE J. Lightwave Technol.*, Vol. 4, No. 9, September 1986, pp. 1334-1351.

[4] Shafir E. and M. Tur, "Phase-Induced Intensity Noise in an Incoherent Fabry-Perot Interferometer and Other Recirculating Devices," *J. Opt. Soc. Am. A*, Vol. 4, No. 1, January 1987, pp. 77-81.

[5] Tur M., E. Shafir, and K. Bløtekjaer, "Source-Induced Noise in Optical Systems Driven by Low-Coherence Sources," *IEEE J. Lightwave Technol.*, Vol. 8, No. 2, February 1990, pp. 183-189.

[6] Weissman, Y., "Analysis of Optical Radio-Frequency Noise Produced by a Linearly Filtered Gaussian Source," *J. Opt. Soc. Am.*, Vol. 7, No. 1, January 1990, pp. 127-133.

[7] Tur M., B. Moslehi, and J. Goodman, "Theory of Laser Phase Noise in Recirculating Fiber-Optic Delay Lines," *IEEE J. Lightwave Technol.*, Vol. LT-3, No. 1, February 1985, pp. 20-30.

Chapter 12
Signal Analysis in Networks That Are Periodic in Time

Components that vary periodically in time (modulators) play an important role in many optical networks. The methodology for the analysis of networks containing modulators was developed in Chapter 6. Here, we develop the methodology for the analysis of the output signals from such networks.

The effect of a time-dependent network on a stationary signal is fundamentally different from the effect of a time-independent network that we considered so far. In particular, the output field of a time-dependent network is generally nonstationary, and may not possess a well defined spectrum. It appears that no general tools can be offered for the analysis of the output signals of networks with an arbitrary time dependence. However, the case of the time-periodic networks is a notable exception.

The output field of a time-periodic network coupled to a stationary source is cyclostationary. Generally, a cyclostationary field may be represented in terms of an infinite set of mutually stationary processes [1]. This implies that it is possible to "decompose" the time-periodic problem into a set of time-independent problems. This fact is manifested in both the output field and its intensity power spectra. These spectra are composed of a series of spectral components (which will also be referred to as *lines* or *sidebands*), each of which may be regarded as representing the output of a certain time-independent network. The central frequencies of these features are displaced with respect to each other by a whole multiple of the modulation frequency. A time-periodic network generates new spectral amplitudes; the width of the output spectra is generally larger than the width of the corresponding source spectra.

All the concepts that we have introduced in the time-independent case have corresponding analogs in the time-dependent case. The analog of a single quantity in the time-independent case is a set of corresponding quantities in the time-dependent case. In particular, the Jones matrix $\boldsymbol{J}(\nu)$ and the H matrix of the time-independent case are replaced by a set of Jones matrices $\boldsymbol{j}_n(\nu)$ and a set of H matrices $\boldsymbol{H}_n(\nu_1, \nu_2)$. A significant portion of this chapter is devoted to the derivation of these sets of matrices from the original (cyclic) Jones matrix. We will see that, unlike the time-independent case, $\boldsymbol{H}_n(\nu_1,\nu_2) \neq \boldsymbol{j}^{\dagger}_n(\nu_1)\boldsymbol{j}_n(\nu_2)$, implying that the analogy between the time-dependent and the time-independent cases must be drawn carefully.

In many instances, the signal of interest is contained in the sharp peak in one of the sidebands of the output intensity power spectrum, most commonly the one centered at the modulation frequency. This signal can be extracted by filtering the detector current with an appropriate narrowband filter. An important question relevant to this operation is the magnitude of the relative intensity noise at the sideband of interest. We will derive a simple answer to this question for the common case of the instantaneous modulators. An interesting result of our analysis is the conclusion that this relative intensity noise depends on the ratio between the modulation frequency and the source field linewidth.

The treatment of time-periodic networks that we present in this chapter will not be on the same level of generality as the treatment of the time-independent networks. In particular, we will limit ourselves to the case of a *stationary* source coupled to a time-periodic network. This may seem to be insufficient, since the use of directly (electrically) modulated sources is very common in optical network applications. The field that is generated by such sources is not stationary. However, for many practical purposes we may regard directly modulated sources as if they were modulated by an appropriate external modulator. If we adopt this point of view, we can attach this imaginary modulator to the input port of the network. In this manner, the problem of a directly modulated source coupled to a time-independent network is transformed to a problem of a stationary source coupled to a time-periodic network.

12.1 THE OUTPUT FIELD OF TIME-PERIODIC NETWORKS

As we have mentioned above, a stationary source coupled to a time-periodic network generates a cyclostationary field. It is not surprising that the period of the cyclostationary field is identical to the period of the network. In order to prove that a field is cyclostationary, we have to demonstrate that its correlation functions are invariant under a shift of the time origin by a whole multiple of the period. For the sake of simplicity, we present the proof only for the second-order correlation function. The general proof is based on the same principles.

We start with the general relation between the input and the output Jones vectors (Eq. (2.10)):

$$\mathbf{A}'(t_2) = \int dt_1 \mathbf{J}(t_2,t_1)\mathbf{A}(t_1)$$

Using the definition of the second-order correlation function (8.26), we obtain

$$\mathbf{G}'(t_1,t_2) = \int d^2\tau \mathbf{J}(t_1,\tau_1)\mathbf{G}(\tau_1,\tau_2)\mathbf{J}^\dagger(t_2,\tau_2) \tag{12.1}$$

Let us denote as usual the period with T, and let k be an integer. From (12.1) it follows that

$$\mathbf{G}'(t_1 + kT, t_2 + kT) = \int d^2\tau \mathbf{J}(t_1 + kT,\tau_1)\mathbf{G}(\tau_1,\tau_2)\mathbf{J}^\dagger(t_2 + kT,\tau_2) \tag{12.2}$$

We introduce now two new integration variables:

$$s_1 = \tau_1 - kT$$

and

$$s_2 = \tau_2 - kT$$

In terms of these variables, we write (12.2) as follows:

$$\mathbf{G}'(t_1 + kT, t_2 + kT) = \int d^2s \mathbf{J}(t_1 + kT, s_1 + kT)\mathbf{G}(s_1 + kT, s_2 + kT)\mathbf{J}^\dagger(t_2 + kT, s_2 + kT) \tag{12.3}$$

Because $\mathbf{J}(t_1,t_2)$ is cyclic and $\mathbf{A}(t)$ is stationary, the last equation may be reduced to

$$\mathbf{G}'(t_1 + kT, t_2 + kT) = \int d^2s \mathbf{J}(t_1,s_1)\mathbf{G}(s_1,s_2)\mathbf{J}^\dagger(t_2,s_2) = \mathbf{G}'(t_1,t_2) \tag{12.4}$$

which is exactly what we had to prove.

In (2.24) we introduced the Fourier representation of a cyclic Jones matrix:

$$\mathbf{J}(t_2,t_1) = \sum_n \exp(-2\pi i n f_0 t_2)\mathbf{j}_n(t_2 - t_1)$$

where $f_0 = 1/T$. In view of (12.4), we can write the output Jones vector $\mathbf{A}'(t)$ as

$$\mathbf{A}'(t) = \sum_n \exp(-2\pi i n f_0 t)\mathbf{a}_n(t) \tag{12.5}$$

where

$$\mathbf{a}_n(t_2) = \int dt_1 \mathbf{j}_n(t_2 - t_1)\mathbf{A}(t_1) \tag{12.6}$$

or, in the frequency-domain (2.25):

$$\mathbf{A}'(\nu) = \sum_n a_n(\nu - nf_0) \tag{12.7}$$

where

$$a_n(\nu) = j_n(\nu)A(\nu) \tag{12.8}$$

The processes $\mathbf{a}_n(t)$ are mutually stationary; the proof of this assertion is elementary, and we leave it as an exercise for the reader.

Equation (12.5) implies that the output field of a general time-periodic network may be regarded as a superposition of the output fields of an infinite ensemble of "elementary" networks. Each of these elementary networks is composed of a time-invariant Jones matrix $\mathbf{j}_n(t_2 - t_1)$ and an ideal (instantaneous) frequency shifter connected in series. The Jones matrix of the frequency shifter that is connected to the nth network is given by $\exp(-2\pi i n f_0 t_2)\delta(t_2 - t_1)\boldsymbol{I}$. This point of view allows us to draw a general signal flow graph, which may be used to represent any time-periodic network by time-invariant transmissions combined in series with appropriate frequency shifters. Such a signal flow graph is portrayed in Figure 12.1.

12.2 THE POWER SPECTRUM OF THE OUTPUT FIELD

12.2.1 The General Formulation

The single-argument correlation matrix $\mathbf{R}'(\tau)$ of the output field is computed with the help of (8.12). Using (12.5), we get the following expression for the coherency matrix of the output field:

$$\mathbf{G}'(t + \tau, t) = \sum_{n_1, n_2} \exp\{2\pi i f_0[n_2 t - n_1(t + \tau)]\}\mathbf{r}_{n_1 n_2}(\tau) \tag{12.9}$$

where

$$\mathbf{r}_{n_1 n_2}(\tau) = \langle \mathbf{a}_{n_1}(t + \tau)\mathbf{a}_{n_2}^\dagger(t)\rangle \tag{12.10}$$

is the (single-argument) cross correlation function of the processes $\mathbf{a}_{n_1}(t)$ and $\mathbf{a}_{n_2}(t)$.

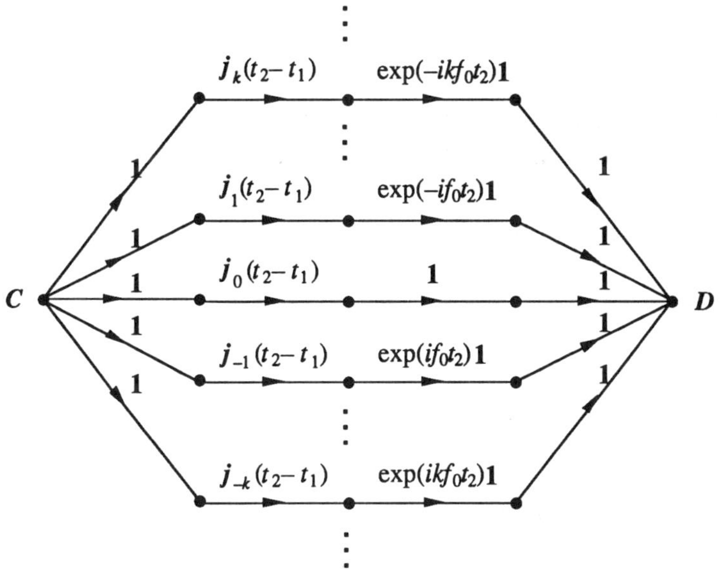

Figure 12.1 A representation of a time-periodic network in terms of its Fourier expansion coefficients (which may be regarded as Jones matrices of time-independent networks) combined with appropriate frequency shifters. The symbol '1' stands for the unity transmission $\delta(t_2 - t_1)\boldsymbol{I}$. The unity branches were added for clarity.

According to (8.12), $\mathbf{R}'(t)$ is the appropriate average of $\mathbf{G}'(t + \tau, t)$:

$$\mathbf{R}'(\tau) = \frac{1}{T}\int_0^T dt\, \mathbf{G}'(t + \tau, t) \tag{12.11}$$

Evaluating this integral, we obtain

$$\mathbf{R}'(\tau) = \sum_n \exp(-2\pi i n f_0 \tau)\mathbf{r}_{nn}(\tau) \tag{12.12}$$

To compute the output field power spectrum $P'(\nu)$, we simply take the Fourier transform and the trace of this equation to get

$$P'(\nu) = \sum_n p_n(\nu - nf_0) \tag{12.13}$$

where

$$p_n(\nu) = \text{Tr}[\boldsymbol{r}_{nn}(\nu)] \tag{12.14}$$

The remaining task is to express $\boldsymbol{r}_{nn}(\nu)$ in terms of the input field correlation matrix $\boldsymbol{R}(\nu)$. This is done with the help of (12.8):

$$\boldsymbol{r}_{nm}(\nu) = \boldsymbol{j}_n(\nu)\boldsymbol{R}(\nu)\boldsymbol{j}_m^{\dagger}(\nu) \tag{12.15}$$

Consequently,

$$p_n(\nu) = \text{Tr}[\boldsymbol{j}_n^{\dagger}(\nu)\boldsymbol{j}_n(\nu)\boldsymbol{R}(\nu)] \tag{12.16}$$

Defining

$$\boldsymbol{h}_{n_1 n_2}(\nu_1, \nu_2) = \boldsymbol{j}_{n_1}^{\dagger}(\nu_1)\boldsymbol{j}_{n_2}(\nu_2) \tag{12.17}$$

We can write (12.16) in a form similar to (10.3):

$$p_n(\nu) = \text{Tr}[\boldsymbol{h}_{nn}(\nu, \nu)\boldsymbol{R}(\nu)] \tag{12.18}$$

12.2.2 Qualitative Features of the Output Field Power Spectrum

It is seen that this spectrum is a superposition of an infinite sequence of spectra $p_n(\nu)$, successively shifted with respect to one another by f_0. The spectra $p_n(\nu)$ are the power spectra of the output fields corresponding to time-independent networks with Jones matrices $\boldsymbol{j}_n(\nu)$. The parameter f_c/f_0 determines the qualitative shape of the output field power spectrum. In the limiting case $f_c \ll f_0$, the output field power spectrum consists of a sequence of well-resolved lines, which may be called *sidebands*. In the other extreme, namely, $f_c \gg f_0$, the output spectrum consists of a single, broad line. Generally, the spectral components $p_n(\nu)$ decay with n, and, in practice, all terms in (12.13) with $|n| > M$ for a certain integer M can be neglected. The linewidth of the output field can be estimated as $2Mf_0 + f_c$, where f_c is the linewidth of the input field.

In Figure 12.2 we show qualitatively the output field power spectrum. The spectrum shown in Figure 12.2(a) corresponds to the case of $f_c \ll f_0$, and, accordingly, it consists of well-resolved sidebands. On the other hand, the spectrum shown in Figure 12.2(b) corresponds to the other extreme, namely, $f_c \gg f_0$. In this limit the sidebands strongly overlap, producing a single, broad line.

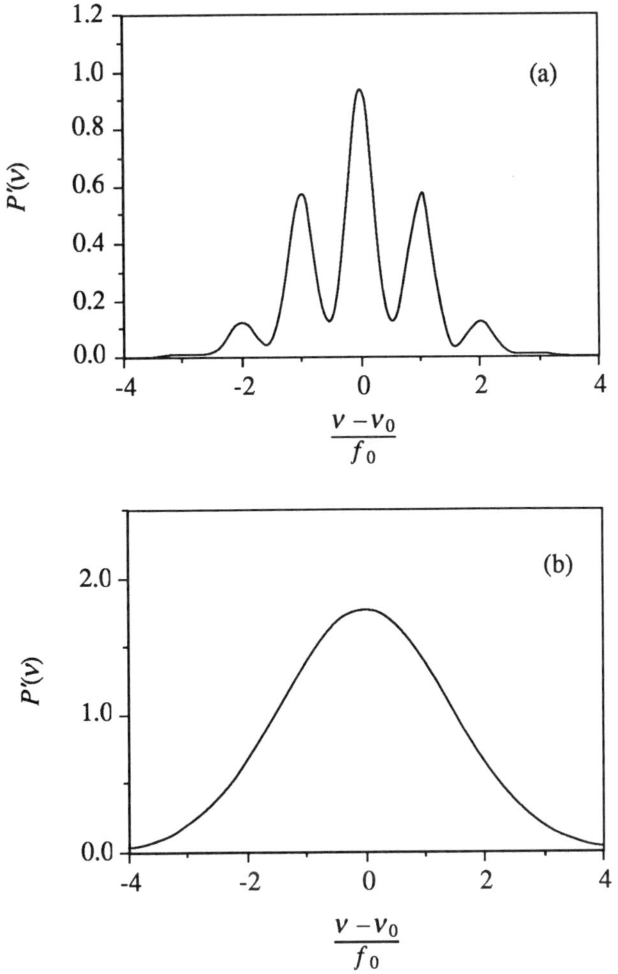

Figure 12.2 The qualitative shape of the output field power spectrum: (a) a series of well resolved lines corresponding to the case in which the modulation frequency is larger than the source linewidth; (b) a single broad line, corresponding to the case in which the modulation frequency is smaller than the source linewidth.

In many cases of interest (such as the cases of the amplitude and the phase modulators), we have $\boldsymbol{h}_{nn}(\nu,\nu) = \boldsymbol{h}_{-n,-n}(\nu,\nu)$, resulting in a power spectrum $P'(\nu)$, which is symmetric around the central frequency of the source field power spectrum. The spectrum shown in Figure 12.2 was assumed to possess this symmetry.

12.2.3 The Average of the Output Intensity

In the case of time-periodic networks, the (ensemble) average of the output intensity is a periodic function of time. Using (12.9) and (8.33), we write

$$<\mathrm{I}'(t)> = \sum_m I'_m \exp(-2\pi i m f_0 t) \tag{12.19}$$

where we have introduced

$$I'_m = \sum_n \mathrm{Tr} <\mathbf{a}_n(t)\mathbf{a}_{n-m}^{\dagger}(t)> \tag{12.20}$$

Since the processes $\mathbf{a}_n(t)$ are mutually stationary, the ensemble average appearing on the right-hand side is simply the cross-correlation function $\mathbf{r}_{n,n-m}(t)$ evaluated at $t = 0$. We can therefore write

$$I'_m = \sum_n \int d\nu \mathrm{Tr}[\boldsymbol{r}_{n,n-m}(\nu)] = \int d\nu \mathrm{Tr}[\boldsymbol{H}_m(\nu,\nu)\boldsymbol{R}(\nu)] \tag{12.21}$$

where we have introduced the H matrices

$$\boldsymbol{H}_m(\nu_1,\nu_2) = \sum_n \boldsymbol{j}_{n-m}^{\dagger}(\nu_1)\boldsymbol{j}_n(\nu_2) \tag{12.22}$$

These developments imply that the analog of the H matrix of the time-independent network is in the present case an *array* of matrices $\boldsymbol{H}_m$. In what follows, we will see that this is indeed the case.

It is of interest to consider the time average of $<\mathrm{I}'(t)>$. This doubly averaged (time and ensemble) output intensity is simply the zeroth Fourier coefficient in the Fourier expansion in (12.19), I'_0. By (12.21),

$$I'_0 = \int d\nu \mathrm{Tr}[\boldsymbol{H}_0(\nu,\nu)\boldsymbol{R}(\nu)]$$

This can also be written in the form

$$I'_0 = \sum_n \int d\nu \mathrm{Tr}[\boldsymbol{j}_n^{\dagger}(\nu)\boldsymbol{j}_n(\nu)\mathbf{R}(\nu)] = \sum_n \int d\nu p_n(\nu)$$

By (12.13), the last expression is simply the integral

$$I'_0 = \int d\nu P'(\nu) \tag{12.23}$$

Thus, I'_0 is equal to the area enclosed between the curve $P'(\nu)$ and the frequency axis. We recall that in the time-independent case this area is equal to the average of the output intensity. In the present case, it is equal to the double (time and ensemble) average of this quantity. Accordingly, the matrix $\boldsymbol{H}_0(\nu,\nu)$ can be interpreted as the power transfer matrix of the time-periodic network.

To prevent a possible mistake, we note that, generally,

$$I'_n \neq \int d\nu p_n(\nu)$$

Consequently, I'_n cannot be interpreted as the average power of the nth sideband. In particular, I'_n is not generally real.

12.2.4 Instantaneous Modulators (Example)

To illustrate this new formalism, we will present here a few elementary examples. We start by considering instantaneous modulators, which are characterized by a Jones matrix of the type in (2.26):

$$\mathbf{J}(t_2,t_1) = \mathbf{U}(t_2)\delta(t_2 - t_1)$$

where $\mathbf{U}(t)$ is a matrix function periodic in time. In this case, the Jones matrices $\boldsymbol{j}_n$ are simply the coefficients of the discrete Fourier transform of $\mathbf{U}(t)$ (Eq. (2.27)), and are independent of ν. In this case, we simply have

$$p_n(\nu) = |\boldsymbol{j}_n|^2 P(\nu) \tag{12.24}$$

Consequently, the power spectrum $P'(\nu)$ of the output field takes the form

$$P'(\nu) = \sum_n |\boldsymbol{j}_n|^2 P(\nu - nf_0) \tag{12.25}$$

In the case of the instantaneous modulators, the H matrices are independent of frequency, and are given by

$$\boldsymbol{H}_m = \sum_n \boldsymbol{j}_{n-m}{}^{\dagger}\boldsymbol{j}_n = f_0{}^2 \sum_n \int_0^T d^2t \exp[2\pi i f_0(nt_2 - nt_1 + mt_1)]\mathbf{U}^{\dagger}(t_1)\mathbf{U}(t_2)$$

Using the identity

$$\sum_n e^{2\pi inx} = \sum_n \delta(x-n) \tag{12.26}$$

we obtain

$$\boldsymbol{H}_m = f_0 \int_0^T dt \exp(2\pi imf_0t)|\mathrm{U}(t)|^2 \tag{12.27}$$

Frequency shifters and phase modulators (Eqs. (2.29), (2.31)) are characterized by

$$|\mathrm{U}(t)|^2 = |\boldsymbol{U}_0|^2$$

Therefore, for such modulators, the only nonvanishing H_m matrix is the one corresponding to $m = 0$. This could be expected, as pure phase modulation does not affect the optical intensity.

Let us now consider two concrete examples. For a polarization-degenerate amplitude modulator, we take (2.28):

$$\mathrm{U}(t) = \frac{1}{2}[1 + \cos(2\pi f_0 t)]\boldsymbol{I} \tag{12.28}$$

For this component, the associated set j_n contains only three nonvanishing members:

$$j_0 = \frac{1}{2}$$

and

$$j_1 = j_{-1} = \frac{1}{4}$$

Consequently,

$$P'(\nu) = \frac{1}{4}P(\nu) + \frac{1}{16}[P(\nu + f_0) + P(\nu - f_0)]$$

Let us consider now the phase modulator. From (2.34) we obtain (taking $\boldsymbol{U}_0 = \boldsymbol{I}$)

$$P'(\nu) = \sum_n |J_n(\gamma)|^2 P(\nu - nf_0) \tag{12.29}$$

where γ is the modulation index (2.31), and $J_n(x)$ is the nth order Bessel function (2.32). The sum in (12.29) is, in principle, infinite; in practice, however, all terms with $|n| \geq M$, where M is an integer much larger than γ, can be neglected.

By (12.27), the power transfer matrix $\boldsymbol{H}_0$ of this phase modulator is unity. This fact can also be verified directly from (12.29), recalling the identity

$$\sum_n |J_n(\gamma)|^2 = 1$$

12.3 THE OUTPUT INTENSITY POWER SPECTRUM

The calculation of the output intensity power spectrum is somewhat tedious, but nevertheless completely straightforward. Most of the complexity comes from the necessity to consider multi-index quantities. Fortunately, in the final result the complexity collapses, and a rather plausible expression is derived.

12.3.1 General Formulation

We start by introducing the intensity correlation function $\rho(t + \tau, t)$:

$$\rho(t + \tau, t) = \langle I(t + \tau) I(t) \rangle \tag{12.30}$$

This correlation function depends on both t and τ, and not just on τ, as in the time-independent case. Let us represent $\rho(t + \tau, t)$ in terms of the output field Jones vector $\mathbf{A}'$:

$$\rho(t + \tau, \tau) = \langle \mathrm{Tr}[\mathbf{A}'(t + \tau)\mathbf{A}'^{\dagger}(t + \tau)]\mathrm{Tr}[\mathbf{A}'(t)\mathbf{A}'^{\dagger}(t)] \rangle \tag{12.31}$$

We proceed further, and represent the Jones vector $\mathbf{A}'$ in terms of the Jones vectors $\mathbf{a}_n(t)$ (Eq. (12.5)):

$$\begin{aligned} \rho(t + \tau, \tau) = & \sum_{n_1, n_2, n_3, n_4} \exp\{2\pi i f_0[-n_1(t + \tau) + n_2(t + \tau) - n_3 t + n_4 t]\} \\ & \cdot \langle \mathrm{Tr}[\mathbf{a}_{n_1}(t + \tau)\mathbf{a}_{n_2}^{\dagger}(t + \tau)]\mathrm{Tr}[\mathbf{a}_{n_3}(t)\mathbf{a}_{n_4}^{\dagger}(t)] \rangle \end{aligned} \tag{12.32}$$

According to the general recipe for the calculation of the power spectra of cyclostationary processes, we have to average $\rho(t + \tau, t)$ with respect to t:

$$\{\rho(t+\tau,t)\}_t = \sum_{n_1,n_2,n_3,n_4} \exp[2\pi i f_0 \tau(n_2 - n_1)]$$

$$\cdot <\mathrm{Tr}[\mathbf{a}_{n_1}(t+\tau)\mathbf{a}_{n_2}{}^{\dagger}(t+\tau)]\mathrm{Tr}[\mathbf{a}_{n_3}(t)\mathbf{a}_{n_4}{}^{\dagger}(t)]>\delta_{n_2-n_1+n_3-n_4,0}$$

where we have used the notation introduced in the incoherent limit treatment for averages of periodic functions. We now introduce $q = n_1 - n_2$:

$$\{\rho(t+\tau,t)\}_t = \sum_{q,n_1,n_3} \exp(-2\pi i q f_0 \tau)$$

$$\cdot <\mathrm{Tr}[\mathbf{a}_{n_1}(t+\tau)\mathbf{a}_{n_1-m}{}^{\dagger}(t+\tau)]\mathrm{Tr}[\mathbf{a}_{n_3}(t)\mathbf{a}_{n_3+m}{}^{\dagger}(t)]> \quad (12.33)$$

The output intensity power spectrum $S'(f)$ is given by the Fourier transform of $\{\rho(t+\tau,t)\}_t$ with respect to τ. From (12.33) it is seen that $S'(f)$ can be cast in the form

$$S'(f) = \sum_q S_q(f - qf_0) \quad (12.34)$$

where

$$S'_q(f) = \sum_{n_1,n_3} \int d\tau e^{2\pi i f\tau} <\mathrm{Tr}[\mathbf{a}_{n_1}(t+\tau)\mathbf{a}_{n_1-q}{}^{\dagger}(t+\tau)]\mathrm{Tr}[\mathbf{a}_{n_3}(t)\mathbf{a}_{n_3+q}{}^{\dagger}(t)]>$$

We now express $S'_q(f)$ in terms of $\boldsymbol{a}_n(\nu)$:

$$S'_q(f) = \sum_{n_1,n_3} \int d^4\nu d\tau \exp\{2\pi i[f\tau - \nu_1(t+\tau) + \nu_2(t+\tau) - \nu_3 t + \nu_4 t]\}$$

$$\cdot <\mathrm{Tr}[\boldsymbol{a}_{n_1}(\nu_1)\boldsymbol{a}_{n_1-q}{}^{\dagger}(\nu_2)]\mathrm{Tr}[\boldsymbol{a}_{n_3}(\nu_3)\boldsymbol{a}_{n_3+q}{}^{\dagger}(\nu_4)]> \quad (12.35)$$

Since the processes $\boldsymbol{a}_n(\nu)$ are mutually stationary, the integrand in (12.35) contains a factor of the form $\delta(\nu_1 - \nu_2 + \nu_3 - \nu_4)$ (not shown explicitly). We can therefore write

$$S'_q(f) = \sum_{n_1,n_3} \int d^4\nu d\tau \exp[2\pi i\tau(f - \nu_1 + \nu_2)]$$

$$\cdot <\mathrm{Tr}[\boldsymbol{a}_{n_1}(\nu_1)\boldsymbol{a}_{n_1-q}{}^{\dagger}(\nu_2)]\mathrm{Tr}[\boldsymbol{a}_{n_3}(\nu_3)\boldsymbol{a}_{n_3+q}{}^{\dagger}(\nu_4)]>$$

Performing the integration over τ, we obtain

$$S'_q(f) = \sum_{n_1,n_3} \int d^4\nu \langle \mathrm{Tr}[\boldsymbol{a}_{n_1}(\nu_1)\boldsymbol{a}_{n_1-q}^{\dagger}(\nu_2)]\mathrm{Tr}[\boldsymbol{a}_{n_3}(\nu_3)\boldsymbol{a}_{n_3+q}^{\dagger}(\nu_4)]\rangle \delta(f - \nu_1 + \nu_2)$$

We now express $S'_q(f)$ in terms of the source Jones vector $\boldsymbol{A}(\nu)$:

$$\begin{aligned} S'_q(f) &= \sum_{n_1,n_3} \int d^4\nu \delta(f - \nu_1 + \nu_2) \\ &\cdot \langle \mathrm{Tr}[\boldsymbol{j}_{n_1}(\nu_1)\boldsymbol{A}(\nu_1)\boldsymbol{A}^{\dagger}(\nu_2)\boldsymbol{j}_{n_1-q}^{\dagger}(\nu_2)]\mathrm{Tr}[\boldsymbol{j}_{n_3}(\nu_3)\boldsymbol{A}(\nu_3)\boldsymbol{A}^{\dagger}(\nu_4)\boldsymbol{j}_{n_3+q}^{\dagger}(\nu_4)]\rangle \\ &= \sum_{n_1,n_3} \int d^4\nu \delta(f - \nu_1 + \nu_2) \\ &\cdot \langle \mathrm{Tr}[\boldsymbol{h}_{n_1-q,n_1}(\nu_2,\nu_1)\boldsymbol{A}(\nu_1)\boldsymbol{A}^{\dagger}(\nu_2)]\mathrm{Tr}[\boldsymbol{h}_{n_3+q,n_3}(\nu_4,\nu_3)\boldsymbol{A}(\nu_3)\boldsymbol{A}^{\dagger}(\nu_4))]\rangle \end{aligned} \quad (12.36)$$

We can express the right-hand side in terms of the network H matrices

$$\begin{aligned} S'_q(f) &= \int d^4\nu \delta(f - \nu_1 + \nu_2) \\ &\cdot \langle \mathrm{Tr}[\boldsymbol{H}_q(\nu_2,\nu_1)\boldsymbol{A}(\nu_1)\boldsymbol{A}^{\dagger}(\nu_2)]\mathrm{Tr}[\boldsymbol{H}_q^{\dagger}(\nu_4,\nu_3)\boldsymbol{A}(\nu_3)\boldsymbol{A}^{\dagger}(\nu_4))]\rangle \end{aligned} \quad (12.37)$$

where we have used the fact that

$$\boldsymbol{H}_{-q}(\nu_4,\nu_3) = \boldsymbol{H}_q^{\dagger}(\nu_3,\nu_4)$$

We can now express $S'_q(f)$ in terms of the source field correlation functions:

$$S'_q(f) = \sum_{n,m,k,j} \int d^4\nu H_{q;mn}(\nu_2,\nu_1)H_q^{\dagger}{}_{jk}(\nu_4,\nu_3)G_{nmkj}(\nu_1,\nu_2,\nu_3,\nu_4)\delta(f - \nu_1 + \nu_2) \quad (12.38)$$

In (12.38), $H_{q;mn}$ and $H_q^{\dagger}{}_{mn}$ denote the (m,n) element of the matrices $\boldsymbol{H}_q$ and $\boldsymbol{H}_q^{\dagger}$, respectively. Comparing (12.38) and (10.8), we see that the expression for $S'_q(f)$ is identical to the expression for the output intensity power spectrum in the time-independent case, with $\boldsymbol{H}_q$ playing the role of $\boldsymbol{H}$. Thus, the output intensity power spectrum $S'(f)$ has a structure that is very similar to the structure of the output field power spectrum: it is an infinite superposition of elementary spectra $S'_q(f)$, each one being shifted with respect to the other by f_0. Each of the elementary power spectra $S'_q(f)$ can be regarded as the output intensity spectrum of the given source coupled to a time-independent network with an H matrix equal to $\boldsymbol{H}_q$. For the calculation of the spectra $S'_q(f)$, we can use all the formalism that we have developed for the computation of the output intensity power spectrum in time-independent networks. In particular, we may use the incoherent limit approximation if all the required conditions are met.

By replacing the field correlation functions in (12.38) with their analytic representation (9.20), we obtain

$$S'_q(f) = \sum_{n,m,k,j} \int d^2\nu H_{q;mn}(\nu_1,\nu_1+f)H_q{}^{\dagger}{}_{jk}(\nu_2,\nu_2+f)\Delta R_{nmkj}(\nu_1+f,f,\nu_2+f)$$
$$+ \int d\nu \mathrm{Tr}[\boldsymbol{H}_q{}^{\dagger}(\nu,\nu+f)\boldsymbol{R}(\nu)\boldsymbol{H}_q(\nu,\nu+f)\boldsymbol{R}(\nu+f)] + |I'_q|^2\delta(f) \tag{12.39}$$

From this expression it is seen that each spectral component $S'_q(f)$ consists of a sharp peak with a magnitude of $|I'_q|^2$ superimposed on a smooth background. Analogously to the time-independent case, we refer to this background as the noise $N'_q(f)$ of the qth spectral component:

$$N'_q(f) = \sum_{n,m,k,j} \int d^2\nu H_{q;mn}(\nu_1,\nu_1+f)H_q{}^{\dagger}{}_{jk}(\nu_2,\nu_2+f)\Delta R_{nmkj}(\nu_1+f,f,\nu_2+\mathrm{f})$$
$$+ \int d\nu \mathrm{Tr}[\boldsymbol{H}_q{}^{\dagger}(\nu,\nu+f)\boldsymbol{R}(\nu)\boldsymbol{H}_q(\nu,\nu+f)\boldsymbol{R}(\nu+f)] \tag{12.40}$$

This function is normally a smooth function of f localized around $f = 0$. We define the output intensity noise power spectrum $N'(f)$ as

$$N'(f) = \sum_q N'_q(f - qf_0) \tag{12.41}$$

The computation of $S'(f)$ can often be simplified by the observation that factors of the form $e^{2\pi i\nu\tau}$ in $\boldsymbol{j}_n(\nu)$ can be discarded. Indeed, if

$$\boldsymbol{j}_n(\nu) = e^{2\pi i\nu\tau}\boldsymbol{j}'_n(\nu)$$

then

$$\boldsymbol{H}_m(\nu_1,\nu_2) = \exp[2\pi i\tau(\nu_1 - \nu_2)]\boldsymbol{H}'_m(\nu_1,\nu_2)$$

From (12.39) it can be seen that a network with a set of matrices H_m will produce the same output intensity power spectrum as a network with a set of matrices H'_m. We have met this rule already in the time-independent case. In the time-dependent case, there is an additional simplification rule, which does not have an analog in the time-independent case. This rule states that factors in $\boldsymbol{j}_n$ in the form e^{inx}, where x is a real constant, can be disregarded too. Indeed, if

$$\boldsymbol{j}_n(\nu) = e^{inx}\boldsymbol{j}_n(\nu)$$

then

$$H_m(\nu_1,\nu_2) = e^{imx} H'_m(\nu_1,\nu_2)$$

Again, from (12.39) it follows that the sets H_m and H'_m generate identical output intensity power spectra.

12.3.2 Qualitative Features of the Output Intensity Power Spectrum

The parameter f_c/f_0 determines the qualitative shape of the output intensity power spectrum. In the limiting case $f_c \ll f_0$, the output intensity power spectrum consists of a sequence of well resolved lines. In the other extreme, namely, $f_c \gg f_0$, the output spectrum consists of a single, broad line. Generally, the spectral components $S'_q(f)$ decay with q, and, in practice, all terms in (12.34) with $|n| > M'$ for a certain integer M' can be neglected. The output intensity power spectra for the two limits are shown qualitatively in Figures 12.3(a) and 12.3(b).

In many applications (the modulated fiber-optic gyroscope being a good example), the signal of interest is contained in one of the spectral components. To extract the signal, the detector output is filtered by a narrowband electrical filter. The filtered signal contains the narrow peak of the corresponding spectral component, together with a portion of its background noise. For such applications it is convenient to introduce the relative intensity noise RIN'_q of the qth spectral component:

$$RIN'_q = \frac{N'(qf_0)}{|I'_q|^2} \tag{12.42}$$

This definition differs from the definition (Eq. (9.65)) of the relative intensity noise power spectrum $RIN(f)$ for a stationary field, in that $RIN(f)$ is a function of the frequency, while RIN'_q depends on a discrete variable q. In the limit $f_c/f_0 \to 0$, we have

$$RIN'_q \approx \frac{N'_q(0)}{|I'_q|^2} \tag{12.43}$$

i.e., the main contribution for the background noise of the q-th peak in the intensity power spectrum comes from the intensity noise N'_q of the corresponding spectral component. On the other hand, if $f_c/f_0 \to \infty$, we obtain

$$RIN'_q \approx \frac{1}{|I'_q|^2} \sum_q N'_q(0) \tag{12.44}$$

These arguments imply that RIN'_q depends on f_c/f_0, and with everything else kept constant, RIN'_q will generally increase when f_c/f_0 increases. This behavior is due to the fact that with increasing f_c/f_0 the overlap between the various spectral components in $S'(f)$ increases.

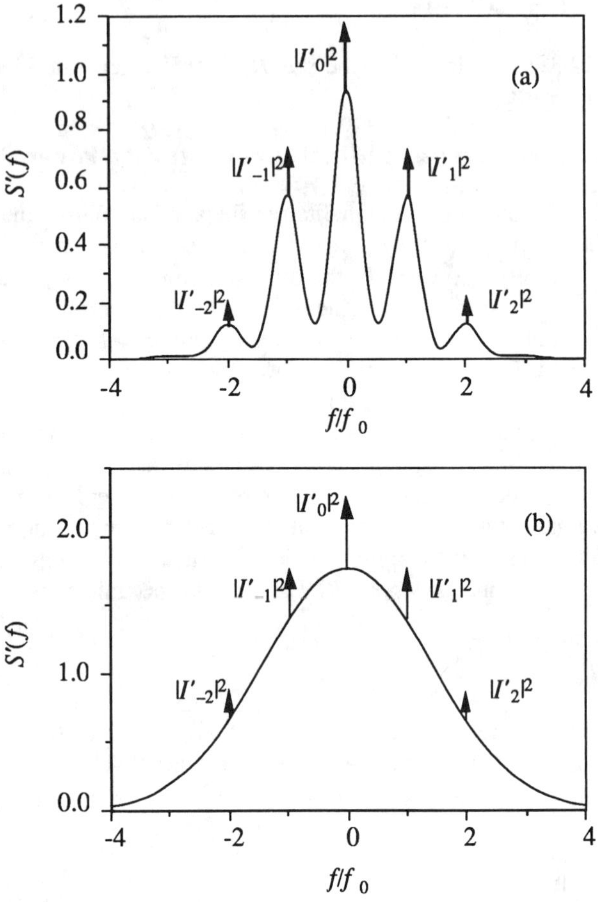

Figure 12.3 Qualitative structure of the output intensity power spectrum: (a) a series of well resolved lines corresponding to the case in which the modulation frequency is greater than the source linewidth; (b) a series of sharp peaks superimposed on a wide background corresponding to the case in which the modulation frequency is smaller than the source linewidth.

In practice, it would be handy to have a simple estimate of RIN'_q for a given source relative intensity noise power spectrum $RIN(f)$. To answer this need, we introduce the *noise figures* NF_q of the modulator:

$$NF_q = \frac{RIN'_q}{RIN(0)} = \frac{<I>^2}{|I'_q|^2} \frac{N'(qf_0)}{N(0)} \tag{12.45}$$

The use of the name "noise figure" in this case is slightly misleading, since generally NF_q depends not only on the modulator properties, but also on the source statistics. However, in the case of the instantaneous modulators discussed below, a simple and general expression for NF_q will be derived. In the limiting cases, the noise factors of instantaneous modulators depend only on the modulator parameters.

12.3.3 Instantaneous Modulators (Example)

Let us reconsider the case of the instantaneous modulators. For simplicity, we assume that the modulator is polarization-degenerate, for which

$$\boldsymbol{H}_q = H_q \boldsymbol{I}$$

For instantaneous modulators, the spectral components $S'_q(f)$ are simply given by

$$S'_q(f) = |H_q|^2 S(f) \tag{12.46}$$

and the output intensity power spectrum takes the form

$$S'(f) = \sum_q |H_q|^2 S(f - qf_0) \tag{12.47}$$

From (12.46) it follows that

$$N'_q(f) = |H_q|^2 N(f) \tag{12.48}$$

and

$$|I'_q|^2 = |H_q|^2 <I>^2 \tag{12.49}$$

For modulators that operate only on the phase, like the frequency modulator or the phase modulator, the only nonvanishing matrix of the set $\boldsymbol{H}_q$ corresponds to $q = 0$. Correspondingly, the output intensity power spectrum for such modulators consists of the $q = 0$ component only. In this case, the output intensity power spectrum is identical, up to a multiplicative constant, to the source intensity power spectrum.

In the case of the instantaneous and polarization degenerate modulators, we can derive a simple expression for the noise figures NF_q. From (12.45) to (12.48) it follows that

$$NF_q = \frac{1}{N(0)|H_q|^2} \sum_n |H_n|^2 N[(q-n)f_0] \tag{12.50}$$

From this equation we derive the two asymptotic values for NF_q:

$$NF_q = \begin{cases} 1 & f_c/f_0 \to 0 \\ \dfrac{1}{|H_q|^2} \sum_n |H_n|^2 & f_c/f_0 \to \infty \end{cases} \tag{12.51}$$

We can write the value of NF_q in the limit $f_c/f_0 \to \infty$ in a more appealing manner. Let us evaluate the time average of $\langle I'(t)\rangle^2$:

$$\{\langle I'(t)\rangle^2\}_t = \sum_{n,m} I'_n I'^*_m \{\exp[2\pi i f_0 t(n-m)]\}_t = \sum_q |H_q|^2 \langle I\rangle^2$$

where we have used (12.49). From this equation and (12.51), we obtain for the limit $f_c/f_0 \to \infty$

$$NF_q = \frac{\{\langle I'(t)\rangle^2\}_t}{|I'_q|^2} \tag{12.52}$$

Let us reconsider the case of the amplitude modulator (12.28). For this case,

$$|U(t)|^2 = \frac{3}{8} + \frac{1}{2}\cos(2\pi f_0 t) + \frac{1}{16}\cos(4\pi f_0 t)$$

From this equation it follows that

$$H_0 = \frac{3}{8}$$

$$H_1 = H_{-1} = \frac{1}{4}$$

$$H_2 = H_{-2} = \frac{1}{16}$$

Consequently, the output intensity power spectrum $S'(f)$ is given by

$$S'(f) = \frac{9}{64} S(f) + \frac{1}{16}[S(f - f_0) + S(f + f_0)] + \frac{1}{256}[S(f - 2f_0) + S(f + 2f_0)]$$

Assuming $N(f)$ in the form

$$N(f) = \frac{<I>^2}{\sqrt{4\pi f_c^2}} \exp\left[-\frac{f^2}{4f_c^2}\right]$$

we may compute the noise factors NF_1, NF_2 from (12.50). The result is shown in Figure 12.4. From (12.51) the asymptotic values of NF_1 and NF_2 as $f_c/f_0 \to \infty$ are 4.375 and 70, respectively, in agreement with Figure 12.4.

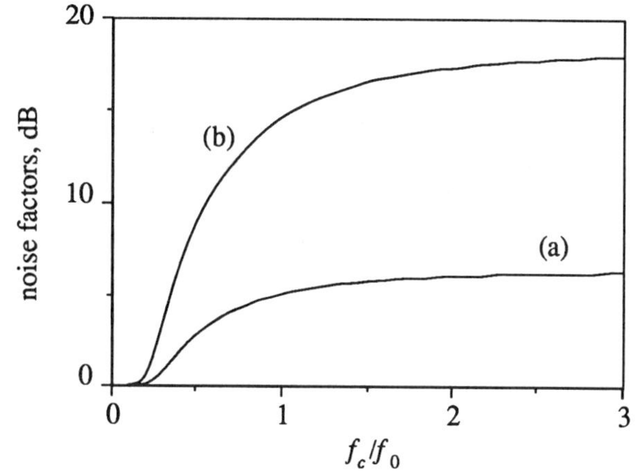

Figure 12.4 The noise factors NF_q for the amplitude modulator as a function of f_c/f_0: (a) NF_1; (b) NF_2.

12.4 ANALYSIS OF THE MODULATED FIBER-OPTIC GYRO

In Chapter 7 we showed that a phase modulator inserted at one of the ends of the Sagnac loop can introduce a bias in the gyro output signal. We now have the appropriate tools to analyze this system properly. We start the analysis with the expression for the Jones matrices $j_k(\nu)$ (Eq. (7.26)):

$$j_k(\nu) = m_k\{r^2\exp[-\pi i(2\nu\tau_S + kf\tau)] - s^2\exp[\pi i(2\nu\tau_S + kf\tau)]\} \tag{12.53}$$

where we have disregarded the factor $e^{\pi i\tau(2\nu+kf)}$. To shorten the notation, we have also introduced r for $\cos\theta$ and s for $\sin\theta$. To remind the reader, m_k are the discrete Fourier expansion coefficients of the modulator, which is assumed to be polarization degenerate.

Before we proceed with the analysis, we introduce an approximation. Normally, the Sagnac phase shifts $2\pi\nu\tau_S$ are much smaller than 1, and therefore $2\pi f_c\tau_S \ll 1$. (This is equivalent to saying that the path difference introduced by the Sagnac effect is much smaller than the source coherence length.) These estimates imply that the variation of $j_k(\nu)$ over the source spectrum linewidth f_c can be neglected. We will therefore approximate $j_k(\nu)$ with

$$j_k \approx m_k\{r^2\exp[-\pi i(2\nu_0\tau_S + kf\tau)] - s^2\exp[\pi i(2\nu_0\tau_S + kf\tau)]\} \tag{12.54}$$

where ν_0 is the central frequency of the source field power spectrum. In this approximation, j_k is independent of ν, as in the case of the instantaneous modulators. From (12.22) we obtain, after some straightforward algebra, the following expression for the corresponding H functions:

$$\begin{aligned} H_m &= [r^4\exp(-\pi iqf_0\tau) + s^4exp(\pi iqf_0\tau)]U_q \\ &\quad - r^2s^2\exp(-\pi iqf_0\tau)[W_q\exp(4\pi i\nu_0\tau_S) + W_{-q}{}^*\exp(-4\pi i\nu_0\tau_S)] \end{aligned} \tag{12.55}$$

where

$$U_q = \sum_n m_{n-q}{}^* m_n \tag{12.56}$$

and

$$W_q = \sum_n \exp(2\pi inf_0\tau)m_{n-q}{}^* m_n \tag{12.57}$$

Let us compute W_q.

$$W_q = f_0{}^2 \sum_n \int_0^T d^2t \exp\{2\pi if_0[-(n-q)t_1 + nt_2 + nt]\}M^*(t_1)M(t_2) \tag{12.58}$$

Using (12.26), we find

$$W_q = f_0 \int_0^T dt \exp(2\pi iqf_0t)M^*(t)M(t-\tau) \tag{12.59}$$

Let us assume now that $M(t)$ is an ideal phase modulator, i.e.,

$$M(t) = \exp[-i\gamma\cos(2\pi f_0 t)] \tag{12.60}$$

We can compute U_q from (12.59) by substituting 0 for τ. For the phase modulator (12.60), we obtain

$$U_q = \delta_{q,0}$$

Using the integral representation (2.32) for the Bessel functions, we get the following result:

$$W_q = J_q(\eta)\exp(-\pi i q f_0 \tau) \tag{12.61}$$

where $\eta = 2\gamma\sin(\pi f_0 \tau)$. Collecting all the above results, we obtain the following expression for the H functions of the modulated fiber-optic gyro:

$$H_q = (r^4 + s^4)\delta_{q,0} - r^2 s^2 J_q(\eta)[\exp(4\pi i \nu_0 \tau_S) + (-1)^q \exp(-4\pi i \nu_0 \tau_S)] \tag{12.62}$$

To take advantage of the modulation scheme, the detector output is filtered by a narrowband filter centered at the modulation frequency f_0. The output signal of this system is proportional to $|I'_1|^2$, where I'_1 is given by

$$I'_1 = H_1 \langle I \rangle = -2ir^2 s^2 J_1(\eta)\sin(4\pi\nu_0\tau_S) \tag{12.63}$$

It is seen that the maximum differential sensitivity occurs at $\tau_S = 0$. The achievement of this fact was the purpose of the modulation scheme. The maximal differential sensitivity is achieved for $r = s = 1/\sqrt{2}$, and for $\eta \approx 1.8$.

There are two general comments that we must make regarding the modulated fiber-optic gyro. First, in most gyro signal processing circuits, both the I'_1 and the I'_{-1} amplitudes are detected and added. Since $|I'_1| = |I'_{-1}|$, our conclusions regarding the sensitivity apply to this detection scheme as well. Secondly, in most gyro systems low coherence sources are used, so that the ratio f_c/f_0 is much larger than 1. Consequently, the noise factors of the gyro system are given by (12.52).

12.5 SUMMARY

In this chapter we have analyzed the problem of a stationary source coupled to a time-periodic network. This problem is of considerable practical importance, since networks containing modulators are very common in many applications.

In the case of periodically modulated networks, the output field and the output intensity power spectra consist of a series of lines which may overlap or be resolved depending upon the ratio between the modulation frequency and the source field linewidth. We have shown that each one of the elementary spectral components may be regarded as an output of a time-independent network. The characteristics of these networks may be derived from the Jones matrix of the original, time-periodic network. To solve the resulting time-independent problems, we can use the methods we developed earlier, including the incoherent limit approximation. To solve the time-periodic problem, we have invoked certain results from the theory of cyclostationary random processes.

The general formulation has been illustrated by the analysis of instantaneous modulators. In spite of its simplicity, this case is of considerable practical importance. In many applications involving modulators, the output signal is filtered with a narrow bandwidth filter centered at one of the output signal sidebands, most commonly the first one. Naturally, there arises the question of the relative intensity noise in such a signal. We have shown that this relative intensity noise depends on the ratio between the modulation frequency and the field linewidth. With everything else kept constant, increasing the modulation frequency normally decreases this relative intensity noise. With an analogy to the electrical network theory, we have introduced the concept of the modulator noise figure, giving the ratio between the input and the output relative intensity noises. Finally we have presented a more complicated application to the case of the modulated fiber-optic gyro.

References

[1] Ogura, H., "Spectral Representation of a Periodic Nonstationary Random Process," *IEEE Trans. Inform. Theory*, Vol. IT-17, March 1971, pp. 143-149.

Chapter 13
Optical Signal and Noise in a Coherent Laser Radar

Readers with a background in optics have probably noted the similarity between single-mode guided systems and classical optical interferometers. In fact, most of the systems we considered in the examples have analogs in unguided optical systems. To emphasize this fact, we have used names taken from classical optical interferometry. The reason for that similarity is that optical interferometers can be regarded in many respects as single-mode optical networks.

Conventional, bulk-optics systems may be regarded as optical networks, with the free space playing the role of the waveguide. When considered as a waveguide, the free space is highly multimode. Therefore, our formalism is not generally suitable for the treatment of such systems. However, interferometric systems may be regarded as an exception to this statement. Interferometric systems contain filters which reduce considerably the number of effective modes. A good example for such a device is the common spatial filter used in many holographic systems. The need to reduce the number of modes stems from the fact that the best interferometer performance in terms of resolution is obtained when all modes higher than the fundamental have been eliminated. The presence of additional modes increases the power throughput of the interferometer, but also decreases its resolution. Consequently, in high-performance interferometric systems, the major contribution to the signal comes exclusively from a single propagation mode (or from two modes, counting polarization). Such systems can be regarded in many respects as single-mode optical networks, and can be treated with our formalism.

In this chapter we present a simple application of the optical signal analysis to the coherent laser radar [1] (lidar). A simplified scheme of such a system is presented in Figure 13.1. The system shown in Figure 13.1 may be regarded as a modulated Mach-Zehnder interferometer. The optical signal reaching the detector is the superposition of two fields, which originate from a common source: one signal is reflected from the target and the second signal is routed directly from the source. The second signal is frequency-shifted and serves as a *local oscillator*. The interference of the two signals on the detector generates a beat, and the receiver is tuned to the beat frequency. This method of detection is called *heterodyne*. Since the beat amplitude is proportional to the local oscillator power, in a heterodyne system the signal is effectively amplified. With a proper local oscillator power, it is possible to reach working conditions in which optical noise mechanisms dominate the noise budget of the system.

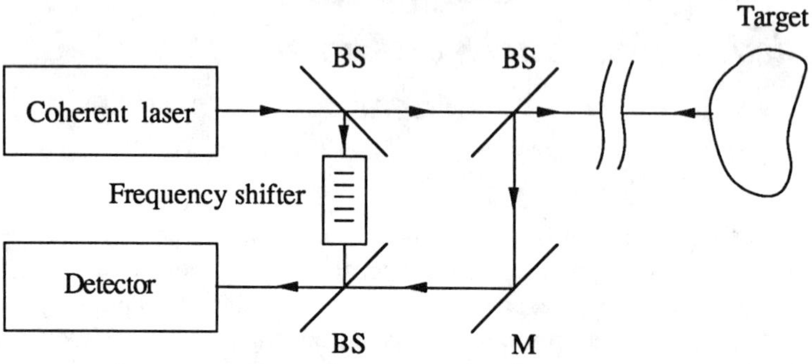

Figure 13.1 A simplified scheme of a coherent lidar. BS: beam splitter, M: mirror.

In this chapter we present an optical signal analysis of the coherent lidar system. To simplify matters, we completely ignore polarization effects, namely we assume that the system is strictly single-mode.

13.1 THE TRANSFER FUNCTIONS OF THE LIDAR SYSTEM

The lidar system contains a frequency shifter, which is a time-periodic component. Therefore, it has to be treated as a time-periodic network, characterized by a cyclic transfer function. As we have seen, the frequency-domain representation of a cyclic transfer function involves (in principle) an infinite set of transfer functions, which we referred to as Fourier expansion coefficients. Our analysis starts with the calculation of these transfer functions.

The modulated Mach-Zehnder system was considered in Chapter 6, and the transfer functions of an instantaneous frequency shifter were considered in Chapter 2. Combining (2.28) and (6.31), we can write the Fourier expansion coefficients of the lidar transfer function as follows:

$$j_k(\nu) = q_0 \exp(2\pi i \nu \tau_1)\delta_{k,0} + q_1 \exp(2\pi i \nu \tau_2)\delta_{k,1} \tag{13.1}$$

where q_0 and q_1 are certain complex constants characteristic of the system, and τ_1 and τ_2 denote the time delays in the two optical paths. Both $|q_0|$ and $|q_1|$ are smaller than 1. The first term in (13.1) represents the signal reflected from the target, and the second term in (13.1) represents the local oscillator signal. Under normal lidar operating conditions the two following relations are satisfied. Since the local oscillator power is much stronger than the power of the signal reflected from the target, we have

$$|q_0| \ll |q_1| \tag{13.2}$$

The distance from the system to the target is normally much larger than the distance between the laser and the detector. Therefore,

$$\tau_1 \gg \tau_2 \tag{13.3}$$

From (13.1) it can be seen that the lidar system has only two nonvanishing transfer functions, j_0 and j_1.

13.2 THE OUTPUT FIELD POWER SPECTRUM

When regarded as an optical network, the output field of the lidar system is the field incident on the detector. Let us compute the power spectrum of this field. Using (12.16) we get

$$p_k = |j_k|^2 P(\nu) = |q_k|^2 P(\nu) \tag{13.4}$$

The output field power spectrum $P'(\nu)$ is derived from (12.13):

$$P'(\nu) = |q_0|^2 P(\nu) + |q_1|^2 P(\nu - f_0) \tag{13.5}$$

where f_0 is the frequency shift. This spectrum is shown schematically in Figure 13.2. The spectrum is not drawn to scale, since, in practice, there are several orders of magnitude difference between the reflected signal and the local oscillator powers.

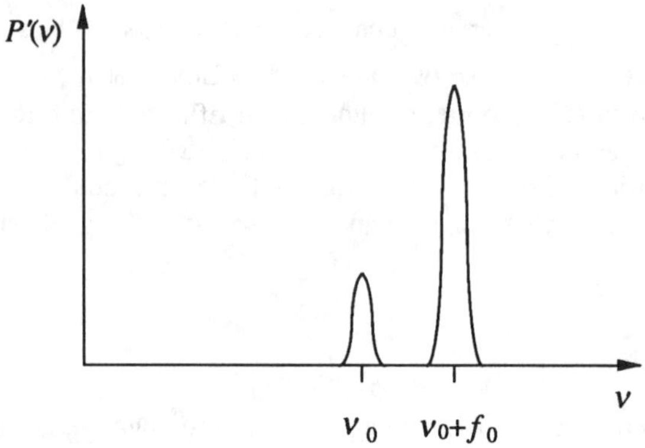

Figure 13.2 The power spectrum of the output field of a lidar system (schematic).

13.3 THE OUTPUT INTENSITY NOISE POWER SPECTRUM

To derive the output intensity noise power spectrum, we will use the general relation (12.40). First, we must calculate the H functions of the system. Combining (12.22) and (13.1), we get

$$H_0(\nu_1,\nu_2) = |q_0|^2\exp[2\pi i\tau_1(\nu_2-\nu_1)] + |q_1|^2\exp[2\pi i\tau_2(\nu_2-\nu_1)] \qquad (13.6)$$

$$H_1(\nu_1,\nu_2) = q_0{}^*q_1\exp[2\pi i(\nu_2\tau_2-\nu_1\tau_1)] \qquad (13.7)$$

and

$$H_{-1}(\nu_1,\nu_2) = q_0q_1{}^*\exp[2\pi i(\nu_2\tau_1-\nu_1\tau_2)] \qquad (13.8)$$

All other H functions vanish. Consequently, the output intensity noise power spectrum consists of only three components. Let us start with the central noise component $N'_0(f)$. From (13.6) we obtain

$$H_0(\nu,\nu+f) = |q_0|^2\exp(2\pi if\tau_1) + |q_1|^2\exp(2\pi if\tau_2) \qquad (13.9)$$

Thus, $H_0(\nu,\nu+f)$ is independent of ν. It follows, therefore, that

$$H_0(\nu_1,\nu_1+f)H_0^*(\nu_2,\nu_2+f) = |q_0|^4 + |q_1|^4 + 2|q_0|^2|q_1|^2\cos(2\pi f\Delta\tau) \quad (13.10)$$

where

$$\Delta\tau = \tau_2 - \tau_1$$

From (12.40) we get

$$N'_0(f) = [|q_0|^4 + |q_1|^4 + 2|q_0|^2|q_1|^2\cos(2\pi f\Delta\tau)]N(f) \quad (13.11)$$

The central noise component $N'_0(f)$ is a product of a frequency-periodic, system-dependent term and the source noise power spectrum. In this respect it is similar to the spectra that are obtained in the incoherent limit of time-independent systems. However, the modulating function in the square brackets is practically flat, in view of (13.2).

We proceed now to the computation of the sideband noise spectrum $N'_1(f)$. This function can also be referred to as the *lineshape* of the heterodyne signal. A simple calculation yields

$$H_l(\nu,\nu+f) = q_0^*q_1\exp[2\pi i(\nu\Delta\tau + f\tau_1)] \quad (13.12)$$

from which it follows that

$$H_1(\nu_1,\nu_1+f)H_1^*(\nu_2,\nu_2+f) = |q_0|^2|q_1|^2\exp[2\pi i\Delta\tau(\nu_1-\nu_2)] \quad (13.13)$$

Similarly, it can be shown that

$$H_{-1}(\nu_1,\nu_1+f)H_{-1}^*(\nu_2,\nu_2+f) = |q_0|^2|q_1|^2\exp[-2\pi i\Delta\tau(\nu_1-\nu_2)] \quad (13.14)$$

The H function products (13.13) and (13.14) are very similar to the product we obtained in the case of the dispersive waveguide. In fact, (13.13) is obtained by substituting $\Delta\tau$ for τ in (10.27), and (13.14) is obtained by substituting $-\Delta\tau$ for τ in the same equation. Therefore, the intensity noise power spectra for the lidar system can be obtained by applying these substitutions in the corresponding expressions we derived earlier for the case of the dispersive waveguide. The analysis of the output intensity noise power spectrum was done for the case of the random-phase source. This case is also appropriate here, since in coherent lidar systems highly coherent sources are used.

The output noise power spectrum for the dispersive waveguide was presented in (10.37). Applying the proper substitutions to that expression, we find that

$$N'_1(f) = \frac{2\gamma P_R P_{LO}}{4\pi^2 f^2 + \gamma^2}\left[1 - \frac{\gamma e^{-\gamma|\Delta\tau|}}{2\pi f}\sin(2\pi f|\Delta\tau|) - e^{-\gamma|\Delta\tau|}\cos(2\pi f|\Delta\tau|)\right] \quad (13.15)$$

where, following (13.5), we have identified $\langle I\rangle|q_0|^2$ and $\langle I\rangle|q_1|^2$ with the received power P_R and the local oscillator power P_{LO}, respectively. Since the expression on the right-hand side depends only on the absolute value of $\Delta\tau$, we have

$$N'_1(f) = N'_{-1}(f) \quad (13.16)$$

In the case of a random-phase source, the central component of the output intensity noise spectrum $N'_0(f)$ vanishes. This is due to the fact that for a random-phase source $N(f) = 0$.

The function $N'_1(f)$ depends on essentially one parameter, $x = \gamma|\Delta\tau|$. To see this, we rewrite (13.15) in a dimensionless form:

$$\gamma N'_1(y) = P_R P_{LO}\,\frac{2x^2}{4\pi^2 y^2 + x^2}\left[1 - x e^{-x}\frac{\sin(2\pi y)}{2\pi y} - e^{-x}\cos(2\pi y)\right] \quad (13.17)$$

where we have also replaced the argument by the dimensionless variable $y = f|\Delta\tau|$. This function is plotted in Figure 13.3.

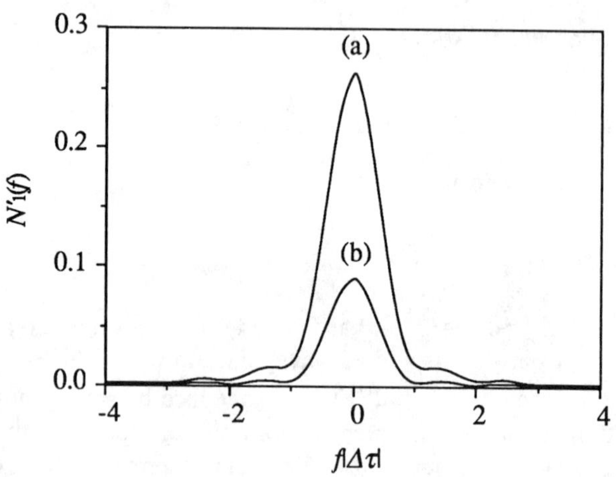

Figure 13.3 The lineshape of a coherent lidar (Eq. (3.17)) normalized to $P_R P_{LO}/\gamma$. (a) $x = 1$, (b) $x = 0.5$.

The output intensity noise power spectrum $N'(f)$ of the lidar system with a random-phase source is given by

$$N'(f) = N'_{-1}(f + f_0) + N'_1(f - f_0)$$

13.4 THE OUTPUT INTENSITY POWER SPECTRUM

The output intensity power spectrum consists of sharp peaks superimposed on the background noise. In the previous section we computed the power spectrum of this background noise. Here, we will compute the amplitudes of these peaks and obtain the complete picture of the output intensity power spectrum.

The amplitude of the mth peak, which is centered at $f = mf_0$ is $|I'_m|^2$, where (Eq. 12.21)

$$I'_m = \int d\nu \mathrm{Tr}[\boldsymbol{H}_m(\nu,\nu)\boldsymbol{R}(\nu)]$$

From (13.12) to (13.14) we get

$$H_0(\nu,\nu) = |q_0|^2 + |q_1|^2 \tag{13.18}$$

$$H_1(\nu,\nu) = q_0{}^*q_1 e^{2\pi i\nu\Delta\tau} \tag{13.19}$$

and

$$H_{-1}(\nu,\nu) = q_0q_1{}^*e^{-2\pi i\nu\Delta\tau} \tag{13.20}$$

Therefore, correspondingly,

$$I'_0 = (|q_0|^2 + |q_1|^2)\langle I\rangle \tag{13.21}$$

$$I'_1 = q_0{}^*q_1R(-\Delta\tau) \tag{13.22}$$

and

$$I'_{-1} = q_0q_1{}^*R(\Delta\tau) \tag{13.23}$$

For a random-phase source (Eqs. (8.70), (8.71)),

$$R(\tau) = \langle I \rangle \exp(-2\pi i \nu_0 \tau - \tfrac{1}{2}\gamma|\tau|) \tag{13.24}$$

Consequently,

$$|I'_0|^2 = (|q_0|^2 + |q_1|^2)^2 \langle I \rangle^2 \tag{13.25}$$

and

$$|I'_1|^2 = |I'_{-1}|^2 = P_R P_{LO} \exp(-\gamma|\Delta\tau|) \tag{13.26}$$

The complete expression for the output intensity power spectrum $S'(f)$ of the lidar system coupled to a random-phase source is

$$\begin{aligned} S'(f) &= P_{LO}^2 \delta(f) + P_R P_{LO} \exp(-\gamma|\Delta\tau|)[\delta(f+f_0) + \delta(f-f_0)] \\ &\quad + N'_{-1}(f+f_0) + N'_1(f-f_0) \end{aligned} \tag{13.27}$$

where we have used the fact that $|q_0| \ll |q_1|$. In practice, the central lobe $N'_0(f)$ of the output intensity noise spectrum does not vanish, since the source always possesses a certain amount of intensity noise. In lidar systems, f_0 is always chosen to be much larger than the source linewidth to prevent overlap between the central and the sideband lobes of the noise spectrum.

13.5 THE SIGNAL-TO-NOISE RATIO IN A COHERENT LIDAR

In this section we will estimate the signal-to-noise ratio in a coherent lidar system, assuming that the dominant contributions to the system noise come from the shot noise and the field-induced noise. To reduce the noise power, the receiver is tuned to two narrow bands of width B surrounding the frequencies f_0 and $-f_0$. The receiver bandwidth is usually determined by the extent of the Doppler shift that has to be accommodated. In most cases, the relevant Doppler shifts are larger than the source linewidth. Therefore, the receiver bandwidth of a coherent lidar designed to detect moving targets is usually larger than γ. If only static targets are of interest, it is possible to reduce B significantly, resulting in improved sensitivity. The bandwidth can also be reduced in Doppler tracking receivers.

Let us write the receiver noise power N as

$$N = N^{(S)} + N^{(F)} \tag{13.28}$$

where $N^{(S)}$ is the contribution of the shot noise, and $N^{(F)}$ is the contribution of the field-induced noise. Since $P_{LO} \gg P_R$, we can write the shot noise equivalent intensity noise density $N^{(S)}_I$ (Eq. (9.68)) as

$$N^{(S)}_I = \frac{q}{\eta} P_{LO} \tag{13.29}$$

It is convenient to define a unit power P_0:

$$P_0 = \frac{\gamma q}{\eta} \tag{13.30}$$

The power P_0 is closely related to the power I_0 defined in (9.71). Physically, these two quantities have the same significance; however, their relative numerical value depends on how exactly the field linewidth f_c is defined. For $\eta = 1$ and $\gamma = 10$ KHz, the value of P_0 is $1.6 \cdot 10^{-14}$ W. The shot noise contribution $N^{(S)}$ to the receiver noise can be written now as

$$N^{(S)} = 2\frac{B}{\gamma} P_0 P_{LO} \tag{13.31}$$

The factor 2 in the right-hand side is present due to the fact that the total receiver bandwidth is $2B$. The signal power P_S is derived from (13.26):

$$P_S = |I'_1|^2 + |I'_{-1}|^2 = 2P_R P_{LO}\, e^{-\gamma|\Delta\tau|} \tag{13.32}$$

Consequently, the following expression for the signal-to-noise ratio SNR is obtained:

$$SNR = \frac{2P_R P_{LO} e^{-\gamma|\Delta\tau|}}{N^{(F)} + 2\frac{B}{\gamma} P_0 P_{LO}} \tag{13.33}$$

In estimating $N^{(F)}$, we will consider two limiting cases. The first case corresponds to the limit $B \ll \gamma$, which is appropriate for a Doppler-tracking system. For this limit, we can estimate

$$N^{(F)} \approx B[N_1(0) + N_{-1}(0)] = 2\frac{B}{\gamma} P_R P_{LO}[1 - (1 + \gamma|\Delta\tau|)e^{-\gamma|\Delta\tau|}] \tag{13.34}$$

so that, in this case, SNR is given by

$$SNR = \frac{e^{-\gamma|\Delta\tau|}}{\frac{B}{\gamma}[1 - (1 + \gamma|\Delta\tau|)e^{-\gamma|\Delta\tau|}] + \frac{B}{\gamma}\frac{P_0}{P_R}} \tag{13.35}$$

For the second limit, $B \gg \gamma$, we estimate

$$N^{(F)} \approx \gamma[N_1(0) + N_{-1}(0)] = 2P_R P_{LO}[1 - (1 + \gamma|\Delta\tau|)e^{-\gamma|\Delta\tau|}] \quad (13.36)$$

and, therefore, SNR in this limit is given by

$$SNR = \frac{e^{-\gamma|\Delta\tau|}}{1 - (1 + \gamma|\Delta\tau|)e^{-\gamma|\Delta\tau|} + \frac{B}{\gamma}\frac{P_0}{P_R}} \quad (13.37)$$

In this case, the signal-to-noise ratio depends on a single system parameter, namely the ratio $BP_0/\gamma P_R$. The behavior of SNR as a function of the product $\gamma|\Delta\tau|$ for three representative values of this system parameter is shown in Figure 13.4.

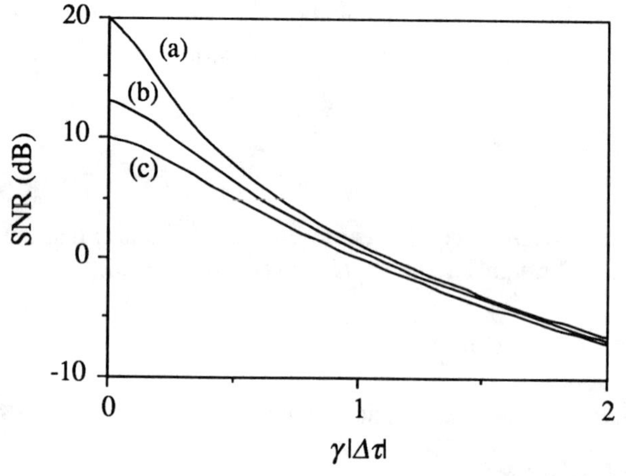

Figure 13.4 The signal-to-noise ratio in a coherent lidar system for a *given* received power P_R (Eq. 13.37). The dimensionless product $\gamma|\Delta\tau|$ is roughly equal to the round-trip distance from the system to the target, divided by the coherence length of the source. The decrease in the signal-to-noise ratio reflects the decrease in phase correlation between the received and local oscillator fields. It is assumed that the receiver bandwidth is much larger than the source linewidth: (a) $BP_0/\gamma P_R = 0.01$; (b) $BP_0/\gamma P_R = 0.05$; (c) $BP_0/\gamma P_R = 0.1$.

The dimensionless product $\gamma|\Delta\tau|$ is roughly equal to the round-trip distance from the lidar to the target divided by the coherence length of the source. From Figure 13.4 it is evident that for $\gamma|\Delta\tau| > 1$ ($\gamma|\Delta\tau| = 1$ corresponds to a range that equals roughly to half the coherence length), an increase in the received power P_R has a negligible effect on the signal-to-noise ratio.

The signal-to-noise ratios calculated from (13.35) and (13.37) depend on the range through the dimensionless product $\gamma|\Delta\tau|$. The decrease of *SNR* with range in (13.35) and (13.37) is caused only by the decrease in the correlation between the received and the local oscillator fields. In evaluating the system performance, one usually has to evaluate the signal-to-noise ratio as a function of range for a *given* transmitter power P_T. Denoting the range by R, we normally have [1]

$$P_R = \frac{\alpha P_T}{R^4} \tag{13.38}$$

where α stands for the appropriate combination of the system parameters and the target cross section. Equation (13.38), combined with (13.35) or (13.37), can be used to evaluate the signal-to-noise ratio of the coherent lidar system as a function of range for a given γ and receiver bandwidth B.

13.6 SUMMARY

In this chapter we have analyzed the optical signal and noise in a coherent laser radar (lidar) system. The lidar is, of course, an unguided system. Nevertheless, it can be regarded as a single-mode optical network and analyzed with the tools developed in this book. The reason for that is the fact that a heterodyne receiver is intrinsically a single-mode system, and the dominant contribution to the output signal comes from the fundamental propagation mode.

There are two main contributions to the coherent lidar receiver noise: the field-induced noise and the shot noise. The output intensity noise power spectrum of the basic coherent lidar consists of three lobes. There is one central lobe centered at $f = 0$, and two sidelobes centered at f_0 and $-f_0$, where f_0 is the amount of the frequency shift used for the heterodyne detection. If the source is ideally a phase-noise source, the central lobe vanishes. The sideband lobes generate field-induced noise in the detector output and deteriorate the signal-to-noise ratio of the system. For a given received and local oscillator powers, the magnitude of the field-induced noise increases with the range to the target. This increase is due to the loss of correlation between the phase of the received field and the local oscillator field. The sidelobe shape was computed for the case of a random-phase source. This model is appropriate for coherent lidar systems, in which stabilized and highly coherent sources are used.

The other important contribution to the system noise budget is the shot noise. The relative magnitude of the shot noise as compared to the field-induced noise depends on the system bandwidth, the source linewidth, the detector responsivity and the range. In typical scenarios, both contributions must be taken into account. Accordingly, we have derived a formula for the system-to-noise ratio which takes into account both the shot noise and the field-induced noise, and can be used for the estimation of the system range. From an inspection of this formula it becomes evident that an increase in the received power has a negligible effect on the signal-to-noise ratio whenever the round-trip distance to the target exceeds the coherence length of the source.

Finally, we should emphasize that, in analyzing real systems, many more aspects must be taken into account. In particular, the performance of terrestrial systems is significantly influenced by atmospheric turbulence. Nevertheless, the relations presented in this chapter can serve as a "zero-order" estimate in most practical cases.

References

[1] Bachman, C.G., *Laser Radar Systems and Techniques*, Dedham, MA: Artech House, 1979.

Index

F

G

H

I

N

O

P

R

S

T

W